Synthesis Lectures on Engineering, Science, and Technology

The focus of this series is general topics, and applications about, and for, engineers and scientists on a wide array of applications, methods and advances. Most titles cover subjects such as professional development, education, and study skills, as well as basic introductory undergraduate material and other topics appropriate for a broader and less technical audience.

Ziad El-Khatib · Sherif Moussa

Wireless Communication Using Deep Learning Techniques for Neuromorphic VLSI Computing

 Springer

Ziad El-Khatib
School of Engineering, Applied Science,
and Technology
Canadian University Dubai
Dubai, UAE

Sherif Moussa
School of Engineering, Applied Science,
and Technology
Canadian University Dubai
Dubai, UAE

ISSN 2690-0300   ISSN 2690-0327 (electronic)
Synthesis Lectures on Engineering, Science, and Technology
ISBN 978-3-031-73799-2   ISBN 978-3-031-73800-5 (eBook)
https://doi.org/10.1007/978-3-031-73800-5

This Springer imprint is published by the registered company Springer Nature Switzerland AG
The registered company address is: Gewerbestrasse 11, 6330 Cham, Switzerland

If disposing of this product, please recycle the paper.

To my beloved family and friends

Contents

Abbreviations and Symbols

AMC	Automatic Modulation Classification
ANN	Artificial Neural Network
AQIF	Adaptive Quadratic Integrate-and-Fire
BPSK	Binary Phase-Shift Keying
CMOS	Complementary Metal Oxide Semiconductor
CNN	Convolutional Neural Network
ConvLSTM	Convolutional Long Short-term
DPI	Differential-Pair Integrator
GRU	Gated Recurrent Unit
HOC	Higher-Order Cumulant
HOM	Higher-Order Moments
LIF	Leaky Integrate-and-Fire
LSTM	Long Short-Term Memory
QIF	Quadratic Integrate-and-Fire
RF	Radio Frequency
RML	Radio Machine Learning
RNN	Recurrent Neural Network
SGD	Stochastic Gradient Descent
SNN	Spiking Neural Network
SNR	Signal-to-Noise Ratio
STDP	Spike-Timing-Dependent Plasticity
VMM	Vector Matrix Multiplication

List of Figures

List of Tables

Introduction

1

1.1 Deep Learning Techniques for AI-Based Wireless Communication Systems

Automatic signal modulation recognition in AI-based wireless communication can be done using combinatorial deep learning neural network techniques to improve resource shortage and spectrum utilization efficiency for dynamic spectrum allocation. Using deep learning neural network circuit methods and doing parallel computations on hardware can reduce costs. Spiking neural network (SNN) provides a promising solution for low-power hardware for neuromorphic computing. Spiking neural network is more promising than other neural networks that can pave a new way for low-power neuromorphic computing applications. Spiking Neural Networks (SNN) is used to connect machine learning and neuroscience. Unlike Artificial Neural Networks (ANN), Spiking Neural Networks (SNN) do not fire continuously. The brain's energy efficiency for decision making cognitive tasks made scientists to focus their efforts on building non-Von Neumann computer systems that imitate the biological brain. Neurons process information as asynchronous event-driven spikes and retain memories as synaptic strengths of their connection in the brain.

Analog VLSI is utilized to design spiking neural networks circuits such as silicon synapse and CMOS neuron. Because transistors have properties similar to nerve membrane channels. When transistors are operated in weak inversion region, they leak a very small current. This transistor region of operation is also known as the subthreshold region. This way a large network of thousands of neurons will consume very low power. Spiking Neural Network (SNN) do not fire continuously. SNN fires only when the post-synaptic potential reaches a certain threshold value making it suitable for low power design.

© The Author(s), under exclusive license to Springer Nature Switzerland AG 2025   1
Z. El-Khatib and S. Moussa, *Wireless Communication Using Deep Learning Techniques for Neuromorphic VLSI Computing*, Synthesis Lectures on Engineering, Science, and Technology, https://doi.org/10.1007/978-3-031-73800-5_1

This chapter presents the introduction. Chapter 2 describes neural network modulation classification. In Chap. 3 describes radio modulation classification optimization using combinatorial deep learning technique. Chapter 4 presents automatic modulation classification performance simulation results. In Chap. 5 Spiking Neural Networks neuromorphic computational system modeling is described. Chapter 6 describes the analysis of a silicon synapse circuit. Synapses are responsible for connecting neurons and communicating spike signals between them. A synapse receives spike voltages from the output of its pre-synaptic neuron. It produces a current based on a weight value. Then it feeds this weighted current to its post-synaptic neuron. In Chap. 6 the design of a fully integrated adaptive modified CMOS synapse circuit is presented. Chapter 6 also presents adaptive CMOS neuron For neuromorphic computing. The design a fully integrated adaptive quadratic integrate-and-fire CMOS neuron was presented. In Chap. 7 the adaptive quadratic integrate-and-fire CMOS neuron performance is presented. Chapter 8 presents the conclusion.

Automatic signal modulation recognition in AI-based wireless communication can be done using combinatorial deep learning techniques to improve resource shortage and spectrum utilization efficiency for dynamic spectrum allocation [1].

A clean signal will have a high Signal-to-Noise Ratio (SNR). We should be able to classify signals in both low and high SNR. Our proposed deep learning model increase accuracy for low SNR and maintain a high classification accuracy for high SNR signals. Using a hybrid deep learning model by combining both ConvLSTM with Transformer-block neural networks shown in Fig. 1.1.

The proposed deep learning modulation classification technique achieves improved classification accuracy of 66% for low SNR signals and 93.5% at high SNR. Thus, getting better accuracy in lower SNR signals without sacrifice accuracy for higher SNR signals. Simulation results show that our proposed deep learning model is robust under noisy signal modulation without the need of denoising the noisy signal.

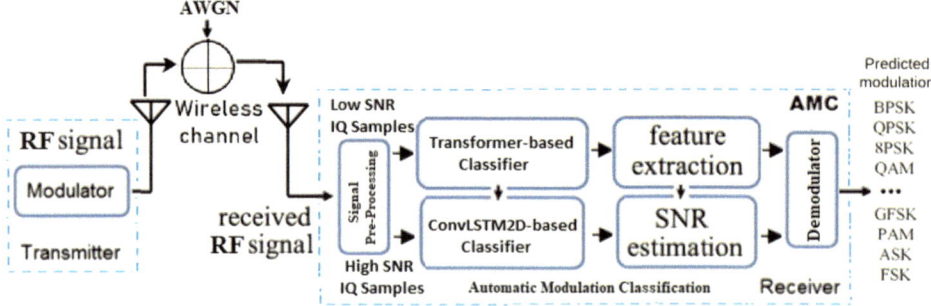

Fig. 1.1 Automatic modulation classification model using combinatorial deep learning technique for both low and high SNR signal classification

By utilizing combinatorial technique in our proposed model that incorporates both ConvLSTM and transformer-block model. Our proposed model has advantage in using transformer-block network since it uses parallelization processing thus make use of parallel computation [2]. Also, the input size can be any size or vector length and simultaneously proceed by the transformer-block network instead of sequentially hence solving the vanishing gradient problem. By using transformer-block network in our proposed model, the context of data can be well captured because it uses the positional encoding and self-attention mechanism which are included in the network modules [2].

Automatic modulation classification (AMC) offers spectrum management and interference detection for software defined radio and cognitive radio networks [1].

As a practical solution to improve the effectiveness of automatic modulation classification, deep learning techniques have been widely used in wireless communication systems. Automatic modulation classification can be divided into likelihood-based and feature-based methods [1]. The likelihood-based method requires more computational complexity. Whereas feature-based method requires less computational complexity [3–6]. Automatic modulation classification model using combinatorial deep learning technique for both low and high SNR signal classification is shown in Fig. 1.1.

The signal is not ideal and is usually combined with noise. A noisy signal will have a low SNR [3–6]. It means that if the noise is higher, the model will likely to fail to do the modulation classification. We should be able to classify signals both in low SNR and high SNR.

The authors West et al. [7] combined CNN and RNN neural networks to form a CLDNN model which improved the modulation recognition accuracy to 85% in high SNR only. The authors Chen et al. [8] combined CNN, RNN, and GAN neural networks to extract the signal spatial characteristics and classification is done with a fully connected layer achieving an accuracy of more than 90% in high SNR only. Jiang et al. [9] combined CLDNN with LSTM neural network achieving 90.8% accuracy at high SNR. Tang et al. [10] combined CNN with GAN achieving an accuracy of 100% at high SNR. Xu et al. [11] combined CNN with LSTM neural networks achieving an accuracy of 90% at high SNR. Jiang et al. [12] combined CNN with Bi-LSTM achieving an accuracy of 93.1%. Liang et al. [13] combined ResNeXt with Attention block achieving an accuracy of 90% in high SNR. Chang et al. [14] combined CNN with Bi-GRU achieving an accuracy of 84% in high SNR only.

Zhang et al. [15] combined GRU with CNN achieving an accuracy of 99.4%. Zou et al. [16] combined CLDNN with Attention achieving an accuracy of 90% in high SNR. The authors Bai et al. [17] combined ResNet with CNN achieving an accuracy of 91% in high SNR. Chen et al. [8] used a deep learning-based attention framework using CNN, RNN, and GAN neural networks. The CNN and RNN are used to extract the signal spatial characteristics. Zou et al. [16] used Attention along with CLDNN neural networks model achieving an accuracy of 90% in high SNR. Bai et al. [17] combined complex value network with ResNet model achieving an accuracy of 91% at high SNR. Chang et al. [14] combined CNN with Bi GRU networks achieving an accuracy of 84% at high SNR. Duan et al. [18]

used combination of CNN with BiLSTM and Attention models achieving an accuracy of 93% at high SNR. Xu et al. [11] combined CNN and LSTM with FC models achieving an accuracy of 90% at high SNR.

Dampage et al. [19] used both LSTM and Bi LSTM models achieving an accuracy of 90% at high SNR. Liu et al. [20] combined DCN and BiLSTM models achieving an accuracy of 90% at high SNR. Yang et al. [21] used an IRS model combined with LSTM neural networks achieving an accuracy of 90% at high SNR. Kumaran et al. [22] combined GRU and BiLSTM neural networks achieving an accuracy of 92% at high SNR. Ze et al. [23] used single neural network model LSTM achieving an accuracy of 90% at high SNR. W. Xie et al used DNN neural network model to extract sixth order cumulant feature of the signal achieving an accuracy of 92% at high SNR. Zhang et al. [24] used BP network model achieving an accuracy of 98% at high SNR. Our proposed deep learning model increase accuracy for low SNR and maintain a high classification accuracy for high SNR signals compared to previous published work. Using a hybrid deep learning model by combining both ConvLSTM with Transformer-block neural networks. Our proposed deep learning modulation classification technique achieves improved classification accuracy of 66% for low SNR signals and 93.5% at high SNR. Other published work such as Oikonomou et al. [25] do not use deep learning models and hence do not have the capability of modulation recognition format prediction also they do not have the capability of loading automatic modulation recognition deep learning model on hardware accelerator chips to take processing load of the main hardware processor compared to our proposed deep learning model [25]. Our proposed automatic modulation recognition format prediction combined deep learning model can be loaded on a hardware accelerator chip thus take the processing load of the main hardware microprocessor. Spiking Neural Networks (SNN) is used to connect machine learning and neuroscience. Unlike Artificial Neural Networks (ANN), Spiking Neural Networks (SNN) do not fire continuously.

Spiking neural network (SNN) provides a promising solution for low-power hardware for neuromorphic computing. Spiking neural network is more promising than other neural networks that can pave a new way for low-power neuromorphic computing applications. Spiking Neural Networks (SNN) is used to connect machine learning and neuroscience. Analog VLSI is utilized to design spiking neural networks circuits such as silicon synapse and CMOS neuron. Because transistors have properties similar to nerve membrane channels. When transistors are operated in weak inversion region, they leak a very small current.

The design of a fully integrated adaptive modified CMOS synapse circuit is presented. By using multiple-gated transistor configuration in the modified CMOS synapse an additional branch provide control where the synaptic output current time-constant is tuned. The effect of changing the multiple-gated transistor bias voltage from 0.25 to 0.45 V tunes the spiking output current exponential time-constant range as shown in simulation results. Our proposed synapse design with multiple-gated transistor configuration achieved a tunable time-constant range of 200 ms compared to previously published work with limited tunable time-constant range to 100 ms. By tuning the decaying exponential time-constant with multiple-gated transistor configuration, the proposed modified CMOS synapse captures the dynamic nature

Table 1.1 ANN-to-SNN learning models Accuracy Performance Comparison

Neuron type	Model type	Network architecture	Recognition accuracy (SSN) (%)
LIF [34]	CNN-to-SNN	Convolutional Neural Network	96
LIF [35]	ANN-to-SNN	3Convn 2Linear	77
LIF [32]	ANN-to-SNN	VGG16	91.5
LIF [33]	ANN-to-SNN	VGG9	90.5

of biological synapses. Merolla [26] synapse design with feedback control achieved a tunable time-constant range is 100 ms. Chen [27] design a memristor-based synapse with tunable time-constant range of 100 ms. Kim [28] synapse design with floating-gate has a 1 ms range of time-constant tuning. Tete [29] synapse design with varying capacitors achieved a tunable time-constant range of 10 us–100 ms. Liu [30] design a memristor-based synapse with tunable time-constant range of 1–100 ms. Hong [31] design a memristor-based synapse with tunable time-constant range of 100 us–100 ms. Table 1.1 shows the tunable of time-constant range of previously published synapse designs in comparison to the proposed synapse design. Our proposed synapse design with multiple-gated transistor configuration achieved a tunable time-constant range of 200 ms compared to previously published work with limited tunable time-constant range to 100 ms.

Moreover, the design of a fully integrated adaptive quadratic integrate-and-fire CMOS neuron was presented as well. A differential pair with variable capacitor integrator and a tunable schmitt trigger threshold detector circuit are integrated in the CMOS neuron that can be tuned varying its spiking frequency. The proposed adaptive quadratic integrate-and-fire CMOS neuron has the ability to adjust the spiking frequency without changing the input current. The simulation results show the proposed adaptive quadratic integrate-and-fire CMOS neuron circuit spiking frequency can be tuned from 58.4 to 312.5 Hz and its spiking period from 17.1 to 3.2 ms with tuning the bias voltage of variable capacitor integrator. Having a peak voltage Vpeak = 0.95 V, a reset voltage Vreset = −0.75 V and a voltage threshold of 0.35 V with a membrane potential range of 1.5 V. The proposed CMOS neuron number of transistors is 26 designed in 130 nm process with a supply voltage of 1.8 V and a total power dissipation of 1.8 mW.

Neuromorphic computing has the potential to be the implementation of choice for low-power Deep Learning systems. Spiking neural networks are regarded as the third generation of artificial neural networks (ANNs). SNNs are more biologically plausible and can be more energy-efficient, especially for applications in neuromorphic computing. Generally, SNNs tend to have lower accuracy compared to their ANN counterparts, but they offer advantages in computational efficiency and energy consumption. ANN-to-SNN refers to the process of converting an Artificial Neural Network (ANN) to a Spiking Neural Network (SNN). ANN-

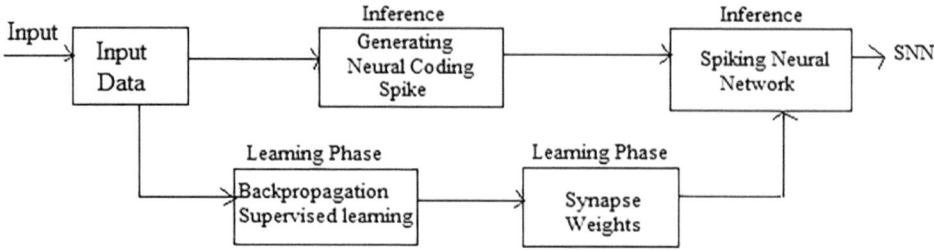

Fig. 1.2 Artificial Neural Networks (ANN) to Spiking Neural Networks (SNN) conversion

to-SNN conversions allows for the efficient and effective training of SNNs by leveraging the knowledge captured in pre-trained ANNs [32–34].

The transformation process from ANN to SNN as shown in Fig. 1.2 starts with ANN is first trained using standard techniques such as backpropagation to learn weights and biases. Then the learned weights are then adjusted or quantized to match the requirements of the SNN framework. During inference, the SNN processes inputs using spikes, and the output is derived based on the spiking activity of the neurons. ANNs operate with continuous values and typically use gradient-based learning rules such as backpropagation. In contrast, SNNs use spikes as their primary means of communication, where neurons fire discrete events based on certain thresholds. The process of converting an ANN to an SNN, which involves adapting learning rules and the representation of information from continuous values to discrete spikes. This transformation involves adapting the continuous activation functions and learning rules of ANNs into the discrete event-driven framework of SNNs [32–34].

In the context of converting an Artificial Neural Network (ANN) to a Spiking Neural Network (SNN), transfer learning plays a crucial role in leveraging the knowledge gained from training the ANN to improve the performance and efficiency of the SNN. Transfer Learning is a technique in machine learning where a model developed for one task is reused as the starting point for a model on a second task. In transfer learning, a pre-trained ANN is used. This model has already been trained on a large dataset, allowing it to capture useful features and representations. The learned weights and biases can be transferred to the SNN, allowing it to start with a good initialization. After transferring the weights, the SNN can be further trained on specific tasks, often using methods like spike-timing dependent plasticity (STDP) or other learning rules designed for spiking networks. First convert the real-valued weights from the ANN to synaptic weights for the SNN, this often involves quantization. Then use a suitable encoding scheme to convert the input data into spike trains. Using STDP to adjust synaptic weights based on the timing of spikes. The key steps involve training the ANN, defining the LIF neuron model, transferring weights, encoding inputs into spike formats, fine-tuning the SNN, then using a pre-trained ANN can significantly reduce the time and data required to train the SNN [32–35].

The learning rules for transforming an ANN to an SNN often involve Spike-Timing Dependent Plasticity (STDP) A biological-inspired learning rule where the strength of connections between neurons is adjusted based on the relative timing of spikes. The weight adaption process where the weights from the ANN are often quantized or adjusted to fit the requirements of the SNN framework. This might involve modifying the continuous weights of the ANN into values suitable for the spiking nature of SNNs. After transferring the weights, the SNN can be further trained on specific tasks, often using methods like spike-timing dependent plasticity (STDP) or other learning rules designed for spiking networks. Table 1.1 shows different ANN-to-SNN Learning Models Accuracy Performance comparison. Using a pre-trained ANN can significantly reduce the time and data required to train the SNN and can be more efficient [32–35].

Deep Learning Techniques for Wireless Communication Systems

2.1 Neural-Network Automatic Modulation Classification

Recent studies show that deep learning models such as neural networks can extract features effectively from various representation of wireless signals such as in-phase and quadrature (IQ) signal or spectrogram in order to achieve high modulation classification accuracy [3–6].

The received signals is preprocessed from I/Q signals cartesian coordinates to polar coordinates to the corresponding amplitude and phase in order to extract more features. By learning more features from the polar domain makes the network more resilient to fading channels [3–6] then are converted to Fig. 2.1 which describes the constellation diagrams of common modulation modes after coordinate transformation.

Figure 2.1 shows various modulation constellation data signals considered in this work that are represented by two vectors one for Quadrature and In-phase representation. Similar modulation constellation data signals are considered in previous publish work Bell et al. [3] where they combined both ResNet and CNN achieving 86% accuracy in high SNR only unlike our proposed work where we achieve high accuracy in both high and low SNR. Also Huang [4] used similar modulation constellation data signals in their published work using compressive CNN achieving accuracy of 95% accuracy in high SNR only. Guo [5] also used similar modulation constellation data signals in their published work using Residual CNN achieving accuracy of 92% accuracy in high SNR only unlike our proposed work where we achieve high accuracy in both high and low SNR.

Zhang et al. [6] proposed a multiscale CNN for constellation-based modulation classification. The network structure was composed of multiple processing modules achieving a classification accuracy can reach 97.7% in high SNR only [36]. Shi et al. [37] and Wang et al. [36] proposed an automatic modulation recognition (AMR) method which includes a

© The Author(s), under exclusive license to Springer Nature Switzerland AG 2025   9
Z. El-Khatib and S. Moussa, *Wireless Communication Using Deep Learning Techniques for Neuromorphic VLSI Computing*, Synthesis Lectures on Engineering, Science, and Technology, https://doi.org/10.1007/978-3-031-73800-5_2

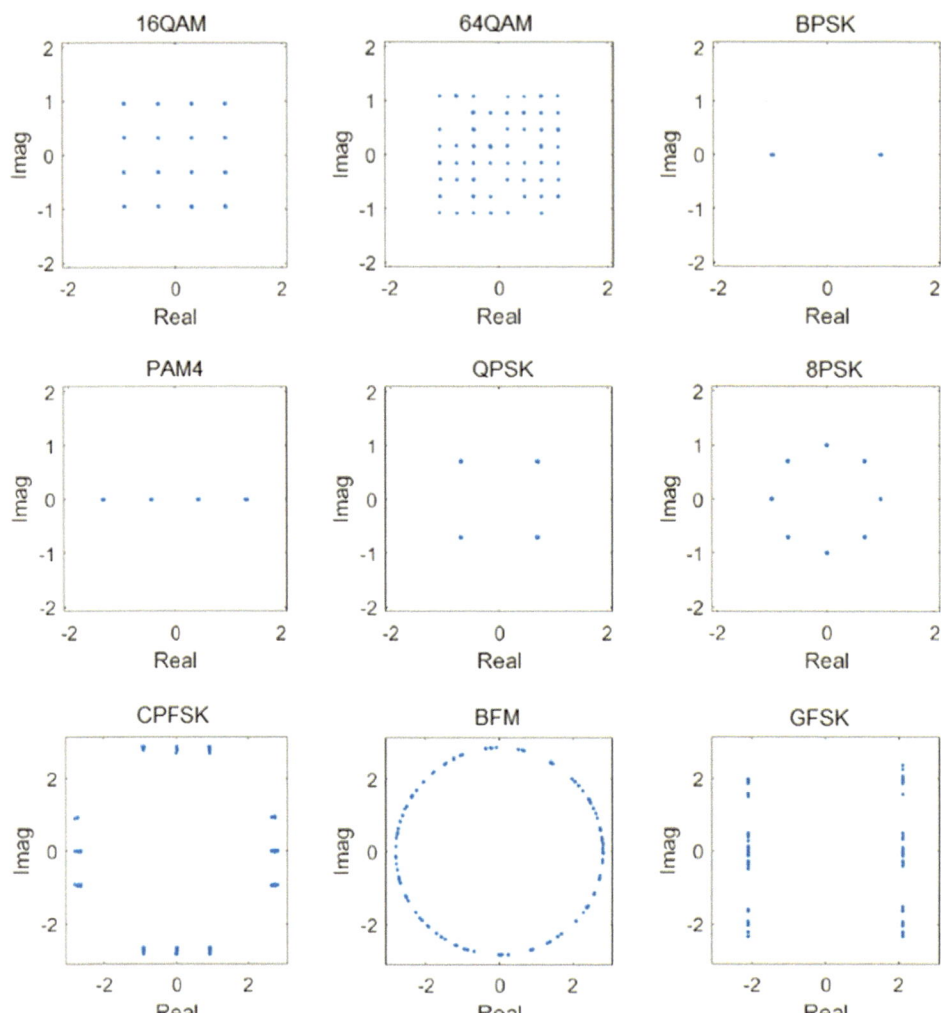

Fig. 2.1 Various Modulation constellation data signals are represented by two vectors one for Quadrature and other for In-phase

multi-scale convolution deep network for recognizing modulation types achieving an overall recognition accuracy of 98.7%.again only in high SNR unlike our proposed work where we achieve high accuracy in both high and low SNR [36, 37].

2.2   Feature Extraction-Based Classification

The proposed deep-learning-based model for automatic modulation classification (AMC) was trained using IQ component signals and image-based constellation diagrams [3–6]. The feature-based method usually requires less computational complexity by extracting data representation features for classification [3–6]. Key features can be categorized as time-domain features including instantaneous amplitude, phase and frequency and frequency-domain features such as wavelet transform of the signals, higher order moments (HOMs) and higher order cumulants (HOCs) that are described in Eqs. 2.2.1 through 2.2.15 as follows [3–6].

The BPSK transmitter emits a signal with voltage of "−a" volts when transmitting bits $X = 0$ and the random variable X is mapped to voltage of "+a" volts when transmitting bits $X = 1$. The communication channel adds the Gaussian noise to this transmitted signal. Therefore, the conditional distribution and variance of σ_n^2 with Gaussian distribution p(r) is the Gaussian distribution with mean "−a" volts variance of σ_n^2 given by and is represented as follows [38]

$$p(r) = \frac{1}{\sqrt{2\pi\sigma_n^2}} \exp\left(-\frac{1}{2}\left(\frac{r+a}{\sigma_n}\right)^2\right) \tag{2.2.1}$$

The ratio when transmitter sends $X = 0$ to $X = 1$ is given by the following λ and is represented as follows [38]

$$\lambda = \frac{\frac{1}{\sqrt{2\pi\sigma_n^2}} \exp\left(-\frac{1}{2}\left(\frac{r-a}{\sigma_n}\right)^2\right)}{\frac{1}{\sqrt{2\pi\sigma_n^2}} \exp\left(-\frac{1}{2}\left(\frac{r+a}{\sigma_n}\right)^2\right)} \tag{2.2.2}$$

which simplifies to the following equation if $\lambda > 1$ the receiver estimates that $X = 1$ is sent else $X = 0$ [38]

$$\lambda = \exp\left(\frac{2ra}{\sigma_n^2}\right) \tag{2.2.3}$$

which is evaluated to

$$\log(\lambda) = \frac{2ra}{\sigma_n^2} \tag{2.2.4}$$

The heterodyne Zero-IF receiver shown in Fig. 2.2 have a decision rule that can be written in a minimum distance which compares the squared Euclidean distances d_1^2 and d_0^2 and is represented as follows [38]

$$d_1^2 - d_0^2 = (r - a)^2 - (r + a)^2 \tag{2.2.5}$$

Various features extracted from IQ signal components, such as amplitude and phase, higher order statistics and higher order cumulants are utilized to provide sequence classification [3–6].

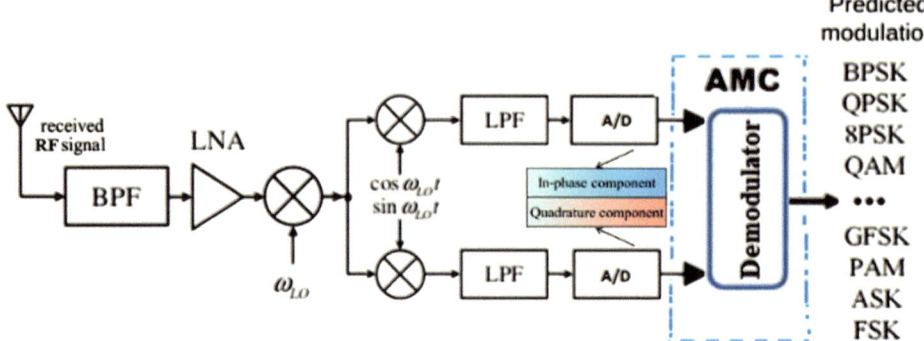

Fig. 2.2 Deep learning-based radio modulation recognition for heterodyne Zero-IF receiver with Quadrature down conversion

A feature extraction-based classification method usually includes two stages data feature extraction and classifier decision making. The key features can be categorized as time domain features including amplitude, frequency, and transform domain features such as higher-order moments (HOMs) and higher-order cumulants (HOCs).

Coefficients series expansion are expected values of complex-valued polynomials H(z). They are computed using the input symbols cross-moments. These input symbols cross-moments used to determine the probability density function the symbols came from. The Gram-Charlier expansion series can be defined as follows [39]

$$f_z(z) = \sum_{p=0}^{\infty} \sum_{q=0}^{\infty} \frac{E\left[H_{p,q}(z)\right]^*}{\sqrt{p!q!}} \frac{H_{p,q}(z)}{\sqrt{p!q!}} \frac{1}{\pi} e^{-zz^*} \qquad (2.2.6)$$

Some complex Hermite polynomials are given by the following coefficients [39]

$$\begin{aligned}
H_{0,0}(z) &= 1 \\
H_{1,0}(z) &= z \\
H_{1,1}(z) &= |z|^2 - 1 \\
H_{2,1}(z) &= z^2 z^4 - 2z \\
H_{2,2}(z) &= |z|^4 - 4|z|^2 + 2
\end{aligned} \qquad (2.2.7)$$

The density functions can be determined by the infinite sequence of these coefficients and can be defined as follows [39]

$$h_{p,q}\left(f_z\right) = \frac{E\left[H_{p,q}(z)\right]^*}{\sqrt{p!q!}} \qquad (2.2.8)$$

The Euclidean distance between two density function or sequence coefficient is determined as follows [39]

$$d\left(f_1, f_2\right) = \sqrt{\sum_{p=0}^{\infty} \sum_{q=0}^{\infty} \left|h_{p,q}\left(f_1\right) - h_{p,q}\left(f_2\right)\right|^2} \qquad (2.2.9)$$

This Euclidean metric on the set of Gram-Charlier coefficients enables use of metric space-based classifiers.

We can define the second-order moments of a stationary random process y(n) with the following coefficients. C_{40}, C_{41}, or C_{42} can be determined in terms of the fourth-and second order moments of y(n) as follows [40, 41]

$$C_{21} = \frac{1}{N} \sum_{n=1}^{N} |y(n)|^2.$$
$$\qquad (2.2.10)$$
$$C_{20} = \frac{1}{N} \sum_{n=1}^{N} y^2(n).$$

The higher order cumulants (HOCs) of fourth order are used for automatic modulation classification. The cumulants of the received symbol y are determined as follows [40, 41]

$$C_{40} = \frac{1}{N} \sum_{n=1}^{N} y^4(n) - 3C_{20}^2$$
$$C_{41} = \frac{1}{N} \sum_{n=1}^{N} y^3(n)y^4(n) - 3C_{20}C_{21} \qquad (2.2.11)$$
$$C_{42} = \frac{1}{N} \sum_{n=1}^{N} |y(n)|^4 - C_{20}^2 - 2C_{21}^2.$$

We can estimate the normalized cumulants as such. By using the above Eqs. 2.2.10 to 2.2.11, three HOC feature parameters (i.e., F_1, F_2, and F_3) are extracted for classification and represented as follows [40, 41]

$$F_1 = \frac{|C_{40}|}{C_{42}}$$
$$F_2 = \frac{|C_{41}|}{C_{42}} \qquad (2.2.12)$$
$$F_3 = \frac{|C_{63}|^2}{|C_{42}|^3}$$

C_40 is utilized to decide whether the constellation diagram is for the real valued rectangular QAM or circular PSK or BPSK or PAM. The amount of cumulative is used to determine the type of modulation. We choose the decision limits as illustrated as follows [40, 41]

$$C_{40} < 0.34 \text{ implies PSK.}$$
$$0.34 < C_{40} < 1.02 \text{ implies QAM.}$$
$$1.02 < C_{40} < 1.68 \text{ implies PAM.} \qquad (2.2.13)$$
$$1.68 < C_{40} < 1.68 \text{ implies BPSK.}$$

Thus, the lowest value of distance between empirical cumulants and theoretical values L tends towards infinity indicates the utilized modulation type.

The signal received by the receiver can be described as follows.

Under the assumption that the signal received by the receiver has undergone carrier synchronization, symbol timing, and matched filtering, and the channel noise is Gaussian white noise, the symbol synchronous sampling complex signal sequence obtained at the output is [42]

$$x(t) = s(t) + n(t) = \sqrt{A} \sum_{k}^{g} \mu_k \sqrt{E_n} \lambda (\text{t}-$$
$$\text{nT}) \exp \lceil j \, (2\pi f_c + \theta_c \rceil + n(t) \qquad (2.2.14)$$

x(t) is the signal received at the receiving end, and s(t) is the signal at the transmitting end, n(t) is the zero-mean complex Gaussian white noise, E_n is signal energy.

Where k = 1,2 ...g, k is index parameter and g is the length of the transmitted code element sequence. A is the unknown factor amplitude, $\lambda(t)$ is transmitted waveform code element; μ_k is code sequence element, θ_c is carrier phase, Ts is width of the code element, E_n is signal energy and f_c is carrier frequency [42].

For zero-mean stationary random process X(t), the p-order mixing moment and k-order HOC are defined as follows.

The characteristic parameters HOCs for various signals that can be designed in accordance is shown in Table 2.1. Utilizing the following equations three feature parameters (i.e., F_1, F_2, and F_3) are extracted [41, 42]

$$F_1 = \frac{|C_{40}|}{C_{42}}$$
$$F_2 = \frac{|C_{41}|}{C_{42}}$$
$$F_3 = \frac{|C_{63}|^2}{|C_{42}|^3} \qquad (2.2.15)$$

The three signal features extracted from HOC having strong low SNR robustness where E_n is signal energy [42].

Table 2.1 High-Order Cumulant HOC modulation signals

Signal	$|C_{40}|$	$|C_{41}|$	$|C_{42}|$	$|C_{63}|$
2ASK	$2E_n^2$	$2E_n^2$	$2E_n^2$	$16E_n^3$
4ASK	$1.36E_n^2$	$1.36E_n^2$	$1.36E_n^2$	$8.32E_n^3$
4PSK	E_n^2	0	E_n^2	$4E_n^2$
2FSK	0	0	0	$4E_n^3$
16QAM	$0.68E_n^2$	$0.68E_n^2$	$0.68E_n^2$	$2.08E_n^3$
64QAM	$0.62E_n^2$	$0.62E_n^2$	$0.62E_n^2$	$1.08E_n^3$

The SNR estimated value can be defined as follows [42]

$$SNR = \frac{\sqrt{2M_2^2 - M_4}}{M_2 - \sqrt{2M_2^2 - M_4}} \qquad (2.2.16)$$

The second-order moment M_2 and fourth order moment M_4 method are given by [42]

$$M_2 = \frac{1}{N} \sum_{n=0}^{N-1} |x(n)|^2$$
$$\qquad (2.2.17)$$
$$M_4 = \frac{1}{N} \sum_{n=0}^{N-1} |x(n)|^4$$

N is the signal length, and the received signal is x(n).

Then we perform normalization to the data by minimizing the penalty function is how Neural networks learn. And accordingly, they iteratively updates a series of weights and biases. Weights in loss function can be used to control the outliers for positive predictions to deal with a class imbalance as shown in Fig. 2.3.

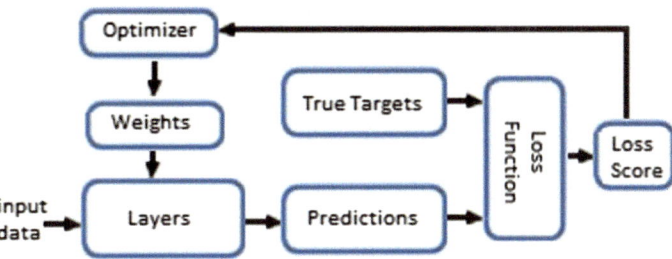

Fig. 2.3 Classification loss function optimization

The weighted cross-entropy loss function can be determined with the following equation. Where yi is actual value of y and y'_i is predicted value of y, w is the weight associated with each sample, M is the total number of samples in the dataset, N is the normalization factor for sample [3, 36, 43]

$$L\left(y'_i, y_i\right) = -\frac{\sum_{i=1}^{M} \omega_i \sum_{i=1}^{M} \frac{y_i}{N_i} \log y'_i}{\sum_{i=1}^{M} \omega_i} \qquad (2.2.18)$$

A categorical cross-entropy loss function use softmax instead of using sigmoid as the last layer activation. The categorical cross-entropy loss function can be determined as follows where M is the total number of samples in the dataset [36, 41, 43]

$$L\left(y', y_i\right) = -\sum_{i=1}^{M} y_i \log y'_i \qquad (2.2.19)$$

In focal loss function has the same softmax of cross-entropy except it is an index of which category is true instead of the target being a probability distribution.

The index of which category is the true value we just pass in as follows where M is the total number of samples in the dataset and γ is the hyperparameter [36, 41, 43].

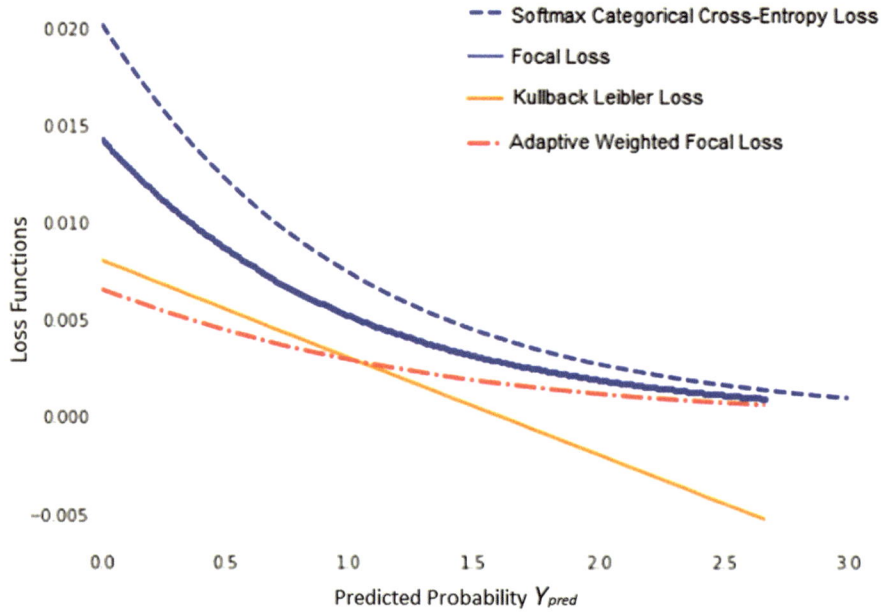

Fig. 2.4 Loss function optimization analysis

$$L\left(y_i', y_i\right) = -\sum_{i=1}^{M} \left(1 - y_i'\right)^{\top} y_i \log y_i' \tag{2.2.20}$$

To avoid underflow issues the Kullback Leibler loss function expects the argument input in the log-space and is computed as follows where M is the total number of samples in the dataset [36, 41, 43]

$$L\left(y_i', y_i\right) = \sum_{i=1}^{M} y_i \left(\log y_i - \log y_i'\right) \tag{2.2.21}$$

An adaptive weighted focal loss is proposed as an optimized loss function for efficient classification which can be used to control the outliers with class imbalance and avoid underflow issues as shown in Fig. 2.4 and it can be determined as follows where Y_{pred} is the predicted probability representing the model's estimated probability that a sample belongs to the positive class [36, 41, 43]

$$L\left(y_i', y_i\right) = -\frac{\sum_{i=1}^{M} \omega_i \sum_{i=1}^{M} \frac{\left(1-y_i'\right)^{\lambda}}{N_i} \left(\log y_i - \log y'_i\right)}{\sum_{i=1}^{M} \omega_i} \tag{2.2.22}$$

Radio Modulation Classification Optimization Using Combinatorial Deep Learning Technique

3

3.1 Combinatorial Deep Learning Techniques for Modulation Classification

In this section, a hybrid combinatorial deep learning model combining both ConvLSTM with Transformer-block neural networks is proposed. Our proposed modulation classifier architecture can learn the signal for both low and high SNR and get better accuracy for signals with high noise. For learning persistent features from a time series data, Recurrent Neural Networks (RNN) are utilized. However, these models using RNNs suffer from much slower training time. LSTM efficient in learning long-term dependencies is a special type of RNN. ConvLSTM Convolutional Long Short-term is a special type of RNN which integrates both CNN with LSTM. ConvLSTM is a modification and extended version of LSTM as shown in Fig. 3.1.

The data transmission and processing in LSTM are realized by three key gate units: input gate, output gate, and forget gate, which are used for implementing information processing.

In LSTM, the input gate i_t, the output gate o_t, and forgotten gates f_t are defined as [19, 20, 44–46], respectively. The equations of LSTM cell are as follows. The input gate and memory status update information are [19, 20, 23, 45].

$$i_t = \sigma \left(W_{xi} X_t + W_{hi} H_{t-1} + W_{ci} C_{t-1} + b_i \right)$$
$$f_t = \sigma \left(W_{xf} X_t + W_{hf} H_{t-1} + W_{cf} C_{t-1} + b_f \right) \tag{3.1.1}$$
$$o_t = \sigma \left(W_{xo} X_t + W_{ho} H_{t-1} + W_{co} C_t + b_o \right)$$

where σ is a sigmoid function and X_t is the input to the current gate structure, and $H_{(t-1)}$ is the output of the previous gate memory cell structure. $C_{(t-1)}$ represent the state of the last

© The Author(s), under exclusive license to Springer Nature Switzerland AG 2025
Z. El-Khatib and S. Moussa, *Wireless Communication Using Deep Learning Techniques for Neuromorphic VLSI Computing*, Synthesis Lectures on Engineering, Science, and Technology, https://doi.org/10.1007/978-3-031-73800-5_3

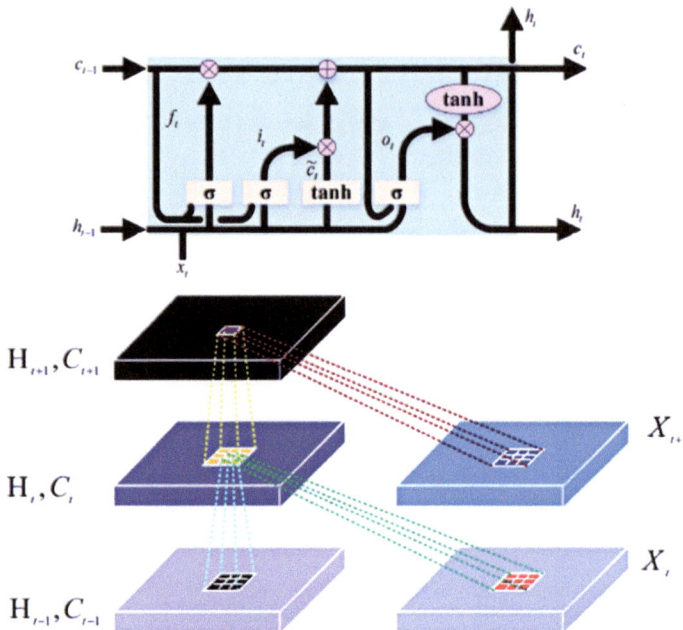

Fig. 3.1 ConvLSTM memory cell structure

memory cell in LSTM. W_{xi} and b_i are the weight and bias of the input gate, W_{xf} and b_f are the weight and bias of the forget gate, and W_{xo} and b_o are the weight and bias of the output gate [19, 20, 23].

The input feature sequence X_t and the output sequence of the previous time $H_{(t-1)}$ are input to the memory cell. The forgetting factor f_t is obtained via the forgetting gate. tanh is an activation function that generates candidate values ($\tilde{C}_t$). In addition, ($\tilde{C}_t$) participates in the calculation to obtain the memory state C_t [19, 20, 23]

$$\tilde{C}_t = \tanh\left(W_{xc}X_t + W_{hc}H_{t-1} + b_c\right).$$
$$c_t = f_t \cdot C_{t-1} + i_t \cdot \tilde{C}_t. \tag{3.1.2}$$
$$h_t = \sigma_t \cdot \tanh\left(c_t\right).$$

The output gate control factor o_t determines whether to output information and is expressed as follows [45]

$$o_t = \sigma\left(W_{xo}X_t + W_{ho}H_{t-1} + W_{co}C_t + b_o\right) \tag{3.1.3}$$

Compared with C_t, $H_{(t-1)}$ contains more information about the current moment. Therefore, $H_{(t-1)}$ represents short-term memory, while C_t represents long-term memory and the state-update is given by the following equations [19, 20, 23].

The Convolutional Long Short-term ConvLSTM shown in Fig. 3.1 calculation equations can be expressed as [47]

$$
\begin{aligned}
i_t &= \sigma \left(W_{xi} X_t + W_{h*} * H_{t-1} + W_{ci} o C_{t-1} + b_i \right). \\
f_t &= \sigma \left(W_{xf} X_t + W_{h**} H_{t-1} + W_{cf} C_{t-1} + b_f \right). \\
o_t &= \sigma \left(W_{xo*} * X_t + W_{ho*} H_{t-1} + W_{co} O C_t + b_o \right). \\
\widetilde{C}_t &= \tanh \left(W_{xc*} X_t + W_{hc*} H_{t-1} + b_c \right). \\
c_t &= f_t o C_{t-1} + i_t o \widetilde{C}_t. \\
H_t &= \sigma_t o \tanh \left(C_t \right).
\end{aligned}
\tag{3.1.4}
$$

Where X_t denotes the input of the current cell, $C_{(t-1)}$ and $H_{(t-1)}$ are state and output of the last cell, respectively. The $*$ operator means the convolution operation and the o denotes the Hadamard product. W denotes the convolution filter. W_{xi} and b_i are the weight and bias of the input gate, W_{xf} and b_f are the weight and bias of the forget gate, and W_{xo} and b_o are the weight and bias of the output gate [47]. ConvLSTM Convolutional Long Short-term contains convolution operation inside it and is used extract spatial-spectral features. It captures long-term and short-term dependencies by stacking multiple ConvLSTM layers [47]. We should be able to classify signals both in low SNR and high SNR. We proposed a combined deep learning architecture shown in the Fig. 3.2 that works to handle the noisy signal modulation low SNR signal modulation and maintain high accuracy for High SNR signals. Adding more ConvLSTM layers with MaxPooling layers can extract more features which is fed into classification layer to predict the probability distribution of each modulation class and is given by the following [3, 48, 49].

$$
P(y = i \mid x, W, b) = \frac{e^{(\omega_{jx} + b_j)}}{\sum_{j=1}^{N} e^{(\omega_j \dot{x} + b_j)}}
\tag{3.1.5}
$$

W is weight and b is the bias of the classification layer with a loss function determined as [48]

$$
L\left(y_i, y_i' \right) = - \sum_{i=1}^{M} \left(y_i \cdot \log y_{y_i'}' \right)
\tag{3.1.6}
$$

Our proposed architecture uses two streams as shown in Fig. 3.1, first stream with ConvLSTM and second stream with transform-block network to extract features for high and low SNR signals. Our proposed model has advantage in using transformer-block network since it uses parallelization processing thus make use of parallel computation [2]. Also, the input size can be any size or vector length and simultaneously proceed by the transformer-block network instead of sequentially hence solving the vanishing gradient problem. By using transformer-block network in our proposed model, the context of data can be well captured because it uses the positional encoding and self-attention mechanism which are included in the network modules [2].

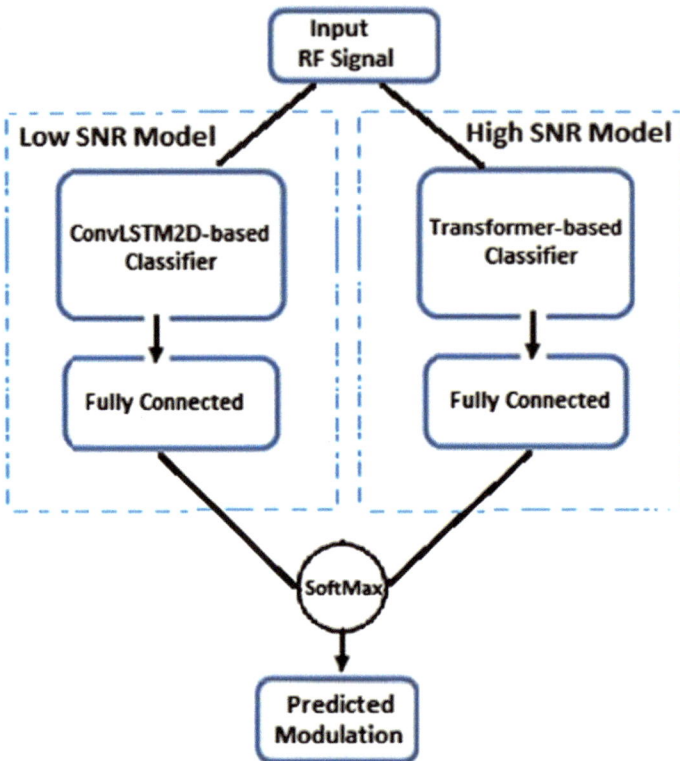

Fig. 3.2 Combinatorial deep learning neural networks model for both low SNR and high SNR modulation classification

The layer type and output shape for ConvLSTM network structure specific to each layer that is implemented is shown in Table 3.1. As illustrated in Table 3.1 a Conv1D convolutional layer consisting of 64 filters is applied to the 128×2 IQ inputs with kernel size of 8.

For higher resolution we need a large filter. It maps the input IQ components onto feature channels. Then, based on the characteristics of IQ signal. Softmax is used as the activation function for multi-class classification and the optimizer used is Adam with loss function categorical cross entropy along with learning rate of 0.01.

After that a MaxPooling1D and then a ConvLSTM and dropout layer follows with the same parameters. After that second ConvLSTM and a dropout layer followed by a Global-AveragePooling1D with the same parameters and a Dense and Dropout layer. At the end is a fully connected layer with an output size 11 with a Softmax activation function.

To increase the overall accuracy performance of our proposed model a ConvLSTM2D model is implemented. The layer type and output shape for ConvLSTM2D network structure specific to each layer that is implemented is shown in Table 3.2. The proposed architecture consists of an input layer of 128×2 IQ input, followed by a Conv2D layer with 64 filters

Table 3.1 ConvLSTM network model layers

Layer type	Output shape
Input	128×2
Conv1D	121×64
MaxPooling 1D	60×64
ConvLSTM	60×64
Dropout	60×64
ConvLSTM	60×64
Dropout	64
GlobalAveragePooling 1D	64
Dense	20
Dropout	20
Dense (Softmax)	11

Table 3.2 ConvLSTM2D network model layers

Layer type	Output shape
Input	128×2
Conv2D	121×64
MaxPooling2D	60×64
ConvLSTM2D	60×64
Dropout	60×64
ConvLSTM2D	60×64
Dropout	64
GlobalAveragePooling2D	64
Dense	20
Dropout	20
Dense (Softmax)	11

and 8 kernels then a MaxPooling2D layer and then a ConvLSTM2D and a Dropout layer follows with the same parameters. After that second ConvLSTM2D and a Dropout layer followed by a GlobalAveragePooling2D with the same parameters and a Dense and Dropout layer. At the end is a fully connected layer with an output size 11. Softmax is used as the activation function for multi-class classification and the optimizer used is Adam with loss function categorical cross entropy along with learning rate of 0.01.

A Transformer-block is utilized as the second stream for larger training data set parallelization [50, 51]. Transformers use attention blocks resulting in faster training time and inference testing time. Our proposed model has advantage in using transformer-block net-

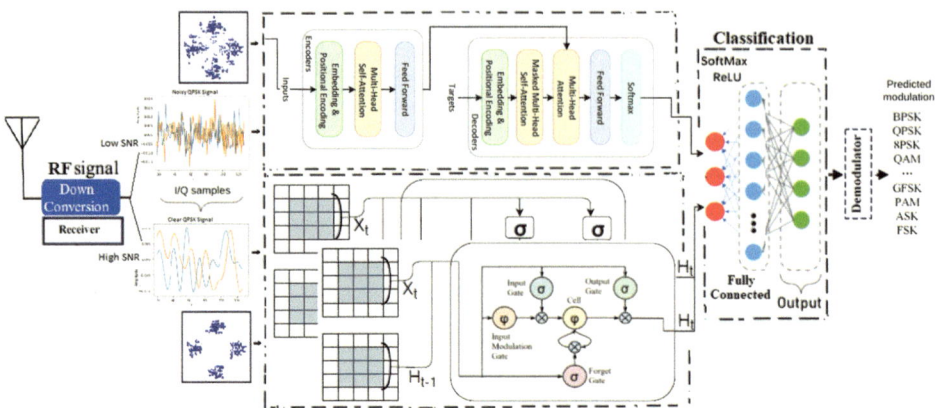

Fig. 3.3 Proposed automatic modulation classification model using combinatorial deep learning technique for both low and high SNR signal classification

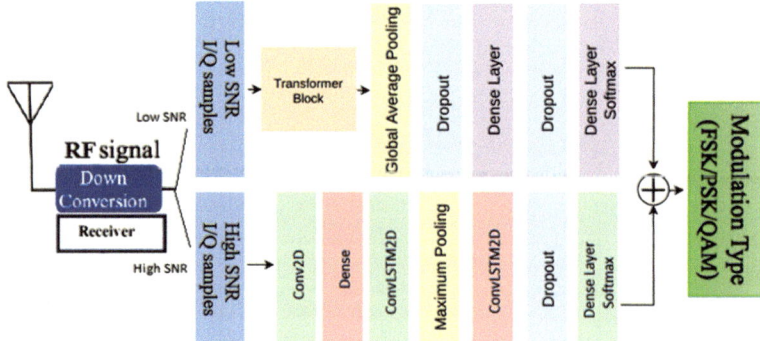

Fig. 3.4 Proposed Combinatorial Deep Learning classification model for both low SNR and high SNR modulation recognition

work since it uses parallelization processing thus make use of parallel computation [2]. The final network design for our proposed classifier model combine Transformer-block with ConvLSTM2D as shown in Figs. 3.3 and 3.4 to achieve higher classification accuracy for both Low and High SNR signals. The proposed multi-stream network can extract various characteristics of signals as shown in Fig. 3.4.

To determine the transformer self-attention, we use the following [2, 52]

$$\text{Attention (Q, K, V)} = \text{softmax}\left(\frac{QK^T}{\sqrt{d_k}}\right)V \qquad (3.1.7)$$

where Q is the query sequence and K is the keys in the sequence and V is the sequence value. And the attention weights are given by [2, 8, 52]

Table 3.3 Transformer-block network model layers

Layer type	Output shape
Input	128×2
Conv1D	121×64
MaxPoolinglD	60×64
ConvLSTM	60×64
Dropout	60×64
Transformer-block	60×64
GlobalAveragePoolinglD	64
Dropout	64
Dense	20
Dropout	20
Dense (Softmax)	11

$$a = \text{softmax}\left(\frac{QK^T}{\sqrt{d_k}}\right) \tag{3.1.8}$$

It uses different transformations activation functions to transform the input and eliminates the need for recurrent connections [2, 8, 13, 52].

Our proposed model has advantage in using transformer-block network since it uses parallelization processing thus make use of parallel computation. In turn faster training time and inference testing time in using parallelization. Moreover, the input size can be any vector length and simultaneously proceed by the transformer-block encoder-decoder attention instead of sequentially hence solving the vanishing gradient problem. Also, context of data is well captured positional encoding and self-attention mechanism.

To increase the overall accuracy performance and to capture more context of the input data, a transformer-block network is implemented as a second stream in our proposed architecture as shown in Fig. 3.4. The layer type and output shape for the transformer-block network structure specific to each layer that is implemented is shown in Table 3.3. The proposed architecture consists of an input layer of 128×2 IQ input, followed by a Conv1D layer with 64 filters and 8 kernels then a MaxPooling1D layer and then a ConvLSTM and a Dropout layer follows with the same parameters. After that a Transformer-block and a GlobalAveragePooling1D followed by Dropout layer with the same parameters and a Dense and Dropout layer. At the end is a fully connected layer with an output size 11. Softmax is used as the activation function for multi-class classification. The Transformer uses a stochastic gradient descent (SGD) optimizer and starts with a learning rate of 0.03, which can be lowered during the training.

3.1.1   RML2016.10a and RML2016.10b Radio Signal Dataset

DeepSig radio signal datasets RadioML2016.10a [53, 54] and RadioML2016.10b [55] are used for evaluating the modulation recognition of our proposed models.

The RML2016.10a dataset parameters [53] are shown in Table 3.4. It contains 220,000 samples, each represented using two vectors each of them has 128 elements Shape: (220,000, 2, 128) [56]. The IQ signal input component. A batch size of 128 is used on each training epoch [54].

The RML2016.10b dataset parameters [55] are shown in Table 3.5. It contains 1.2M samples, each represented using two vectors each of them has 128 elements Shape: (1200000, 2, 128) [56]. The IQ signal input component. A batch size of 128 is used on each training epoch.

Table 3.4 RadioML2016.10a DeepSig dataset parameters

Parameter	Value
Modulations	11 Classes, 8 digital, 3 analog
Length per sample	2×128
Signal format	In-phase & Quadrature IQ
Samples per symbol	8
Total number of samples	220,000
Sampling	1 MHz
Frequency	$[-20 : 2 : 18]$dB
SNR range	176,000
Training samples	44,000
Test samples	

Table 3.5 Table RadioML2016.10b DeepSig dataset parameters

Parameter	Value
Modulations	10 Classes, 8 digital, 2 analog
Length per sample	2×128
Signal format	In-phase & Quadrature IQ
Signal dimension	2×128 per sample
Total number of samples	1,200,000
Duration per sample	128 μs
SNR range	$[-20 : 2 : 18]$dB
Training samples	960,000
Test samples	240,000

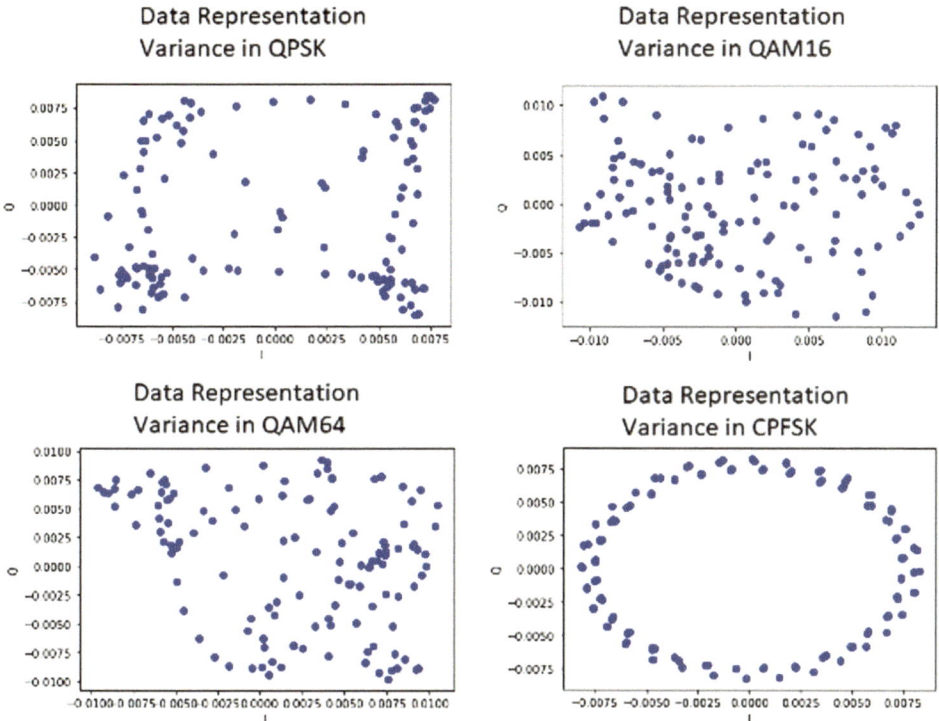

Fig. 3.5 IQ Signal data samples constellation diagram where signals are represented by two vectors one for Quadrature and other for In-phase representation

The Nvidia Tesla A100 Tensor Core GPU is used to speed up the calculation. Models are implemented done using Keras framework with Nvidia A100 Tensor Core GPU.

The 128-sample baseband IQ time-domain signal data is used to identify the modulation type out of 11 modulations based on power spectrum and time. The input data is fed in where the real and imaginary parts of samples are separated as shown in Figs. 3.5 and 3.6 [57].

Received r(t) signal is sampled into its discrete signal r[n], that is of the in-phase (I) components $r^I[n]$ and quadrature (Q) components $r^Q[n]$ are given as [46, 58, 59]

$$r[n] = r^I[n] + jr^Q[n] \tag{3.1.9}$$

In reality the signal is always not ideal and combined with unwanted signal that is considered as noise which can affect our ability to determine the signal. Their relationship with the transmitter side I[n] and Q[n] is given by [46, 59]

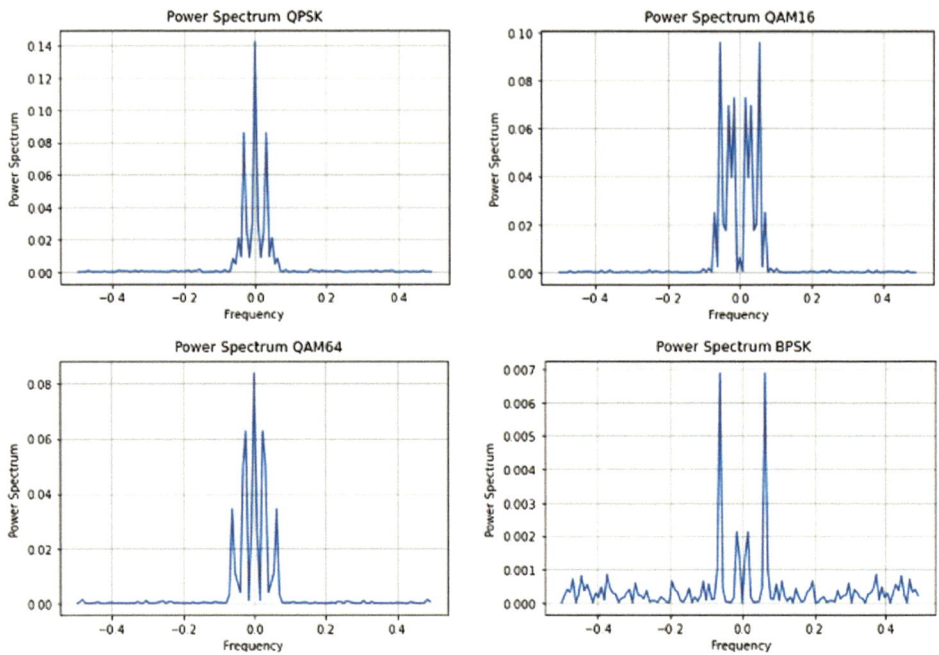

Fig. 3.6 IQ Signal data samples power spectrum representation

$$\frac{r^I[n]}{n_{\text{add}}[n].} = I[n]\cos(2\pi f n + \varphi) - Q[n]\sin(2\pi f n + \varphi) + + \qquad (3.1.10)$$

$$r^Q[n] = -I[n]\sin(2\pi f n + \varphi) - Q[n]\cos(2\pi f n + \varphi) + \atop n_{\text{add}}[n]. \qquad (3.1.11)$$

To construct the relation between I and Q components the transformation function can be expressed as [46]

$$r[n] = \text{radius}[n] = \sqrt{I[n]^2 - Q[n]^2}. \qquad (3.1.12)$$

$$\theta[n] = \text{ theta }[n] = \arctan\left(\frac{I[n]}{Q[n]}\right). \qquad (3.1.13)$$

We detect signals from their time representation showing value of each signal at a given time where each sample is represented by two vectors one for Quadrature and other for In-phase representation and the variance of data shown in Figs. 3.5 and 3.6. We should be able to classify signals both in low SNR and high SNR.

Modulation Classification Performance Simulation Results

4

4.1 Automatic Modulation Classification Performance Simulation Results

The simulation results show that our proposed deep learning model is robust under noisy signal modulation without the need of denoising the noisy signal. The simulation results show our technique outperforms existing feature-based extraction architectures in terms of modulation recognition performance getting better accuracy in lower SNR signals without sacrifice accuracy in higher SNR signals.

Our proposed deep learning modulation classification technique achieves improved classification accuracy of 66% for low SNR signals and 93.5% at high SNR. As shown in Figs. 4.1, 4.2 and 4.3 overall classification accuracy against SNR.

The overall classification performance of our proposed deep learning model for high SNR signals is shown in Fig. 4.1. Two ConvLSTM2D layers gives better accuracy than one ConvLSTM2D layer with batch size of 256 achieving accuracy of 93.5% for High SNR signals.

The overall classification performance of our proposed deep learning model for high SNR signals is shown in Fig. 4.1. Tuning the learning rate hyperparameter to 0.0001 with 200 epochs gives higher classification accuracy achieving accuracy of 93.5% for High SNR signals.

As shown in Fig. 4.2 our proposed deep learning modulation classification technique achieves improved classification accuracy of 66% for low SNR signals and 93.5% at high SNR. Using our hybrid deep learning model by combining both ConvLSTM2D with Transformer-block neural networks, the proposed modulation classifier architecture can learn the signal for both low and high SNR and get better accuracy for signals with high noise. Showing that our model is robust under noisy signal modulation.

We analyze the classification accuracy of our proposed model for different modulation with DeepSig RadioML2016.10a dataset. The confusion matrix presents what modulation

Z. El-Khatib and S. Moussa, *Wireless Communication Using Deep Learning Techniques for Neuromorphic VLSI Computing*, Synthesis Lectures on Engineering, Science, and Technology, https://doi.org/10.1007/978-3-031-73800-5_4

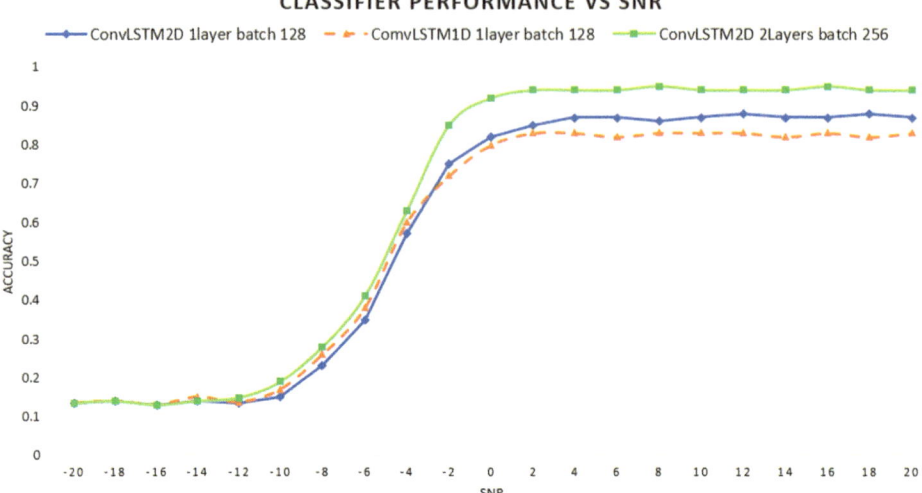

Fig. 4.1 Classifier Accuracy Performance versus SNR for High SNR modulation recognition Increase Model Layer with batch size

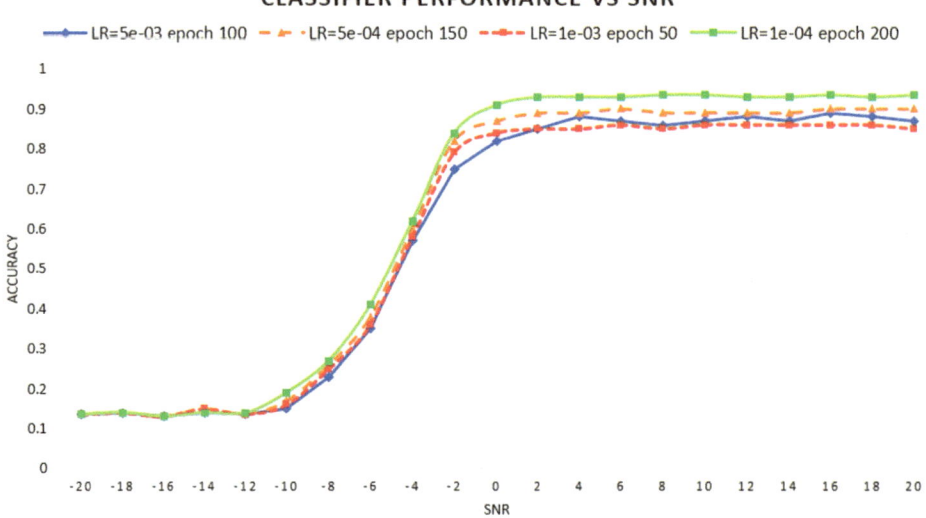

Fig. 4.2 Varying Learning Rate and epoch with hyperparameter tunning

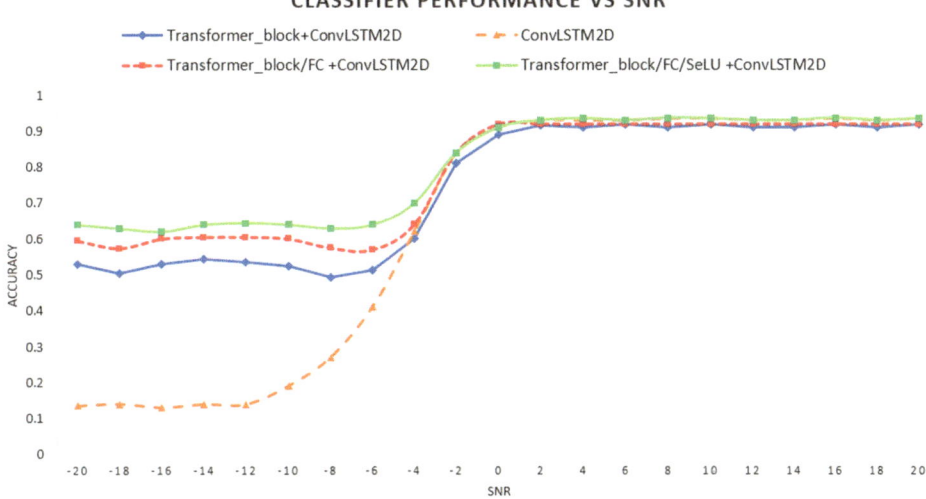

Fig. 4.3 Combined Transformer-block with ConvLSTM Neural Network SNR Classifier Performance

classes the model is confusing with other modulation classes. A dark blue along the diagonal represents a perfect classification.

The confusion matrix performance of the proposed deep learning model for digital modulation signals at 0 dB SNR with 50 epochs is shown in Fig. 4.4. We analyze the classification accuracy of our proposed model for different modulation with DeepSig RadioML2016.10a dataset.

The confusion matrix performance of the proposed deep learning model for digital and analog modulation signals at 0 dB SNR with 300 epochs is shown in Fig. 4.5. We analyze the classification accuracy of our proposed model for different modulation with DeepSig RadioML2016.10a dataset.

The confusion matrix performance of the proposed deep learning model for digital modulation signals at 2 dB SNR with 50 epochs is shown in Fig. 4.6.

The confusion matrix performance of the proposed deep learning model for digital and analog modulation signals at 2 dB SNR with 300 epochs is shown in Fig. 4.7.

The confusion matrix performance of the proposed deep learning model for digital modulation signals at −16 dB SNR with 50 epochs is shown in Fig. 4.8.

The confusion matrix performance of the proposed deep learning model for digital and analog modulation signals at −14 dB SNR with 300 epochs is shown in Fig. 4.9.

The confusion matrix performance of the proposed deep learning model for digital modulation signals at 4 dB SNR with 50 epochs is shown in Fig. 4.10.

The confusion matrix performance of the proposed deep learning model for digital and analog modulation signals at 4 dB SNR with 300 epochs is shown in Fig. 4.11.

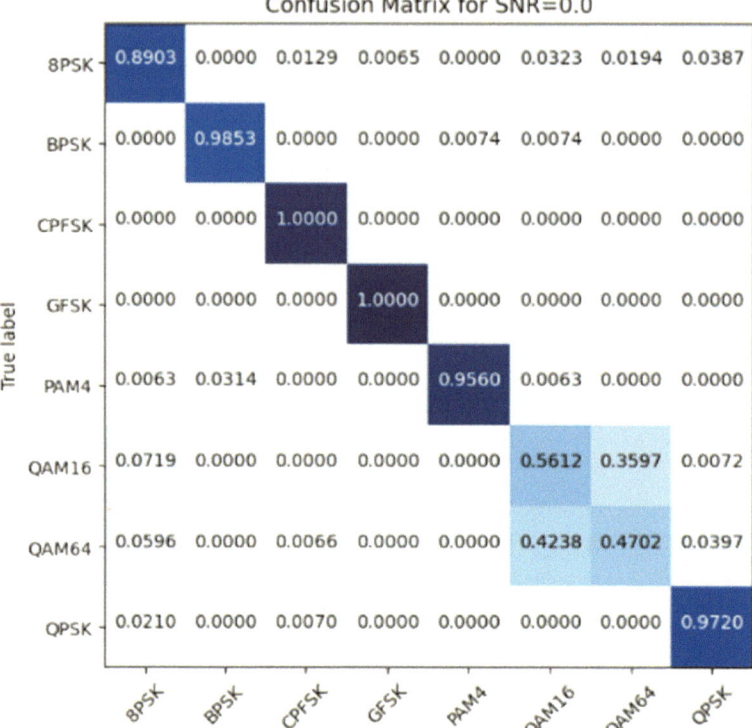

Fig. 4.4 Confusion Matrix of our proposed deep learning model for digital modulations at 0 dB SNR with 50 epochs

The confusion matrix performance of the proposed deep learning model for digital modulation signals at 6 dB SNR with 50 epochs is shown in Fig. 4.12.

The confusion matrix performance of the proposed deep learning model for digital and analog modulation signals at 6 dB SNR with 300 epochs is shown in Fig. 4.13.

The confusion matrix performance of the proposed deep learning model for digital modulation signals at 8 dB SNR with 50 epochs is shown in Fig. 4.14.

The confusion matrix performance of the proposed deep learning model for digital and analog modulation signals at 8 dB SNR with 300 epochs is shown in Fig. 4.15.

The confusion matrix performance of the proposed deep learning model for digital modulation signals at 10 dB SNR with 50 epochs is shown in Fig. 4.16.

The confusion matrix performance of the proposed deep learning model for digital and analog modulation signals at 10 dB SNR with 300 epochs is shown in Fig. 4.17.

The confusion matrix performance of the proposed deep learning mode for digital modulation signals at 12 dB SNR with 50 epochs is shown in Fig. 4.18.

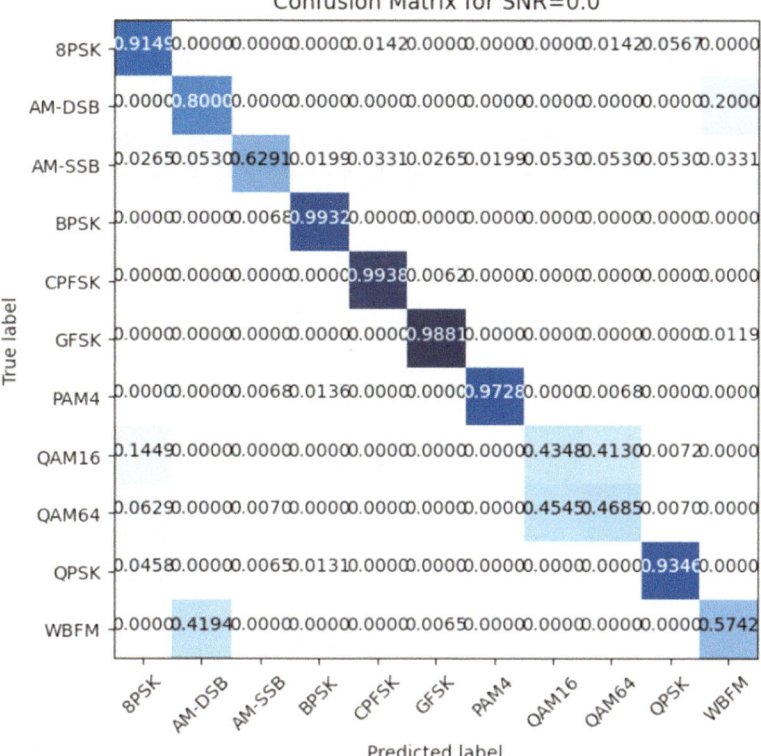

Fig. 4.5 Confusion Matrix of our proposed deep learning model for both digital and analog modulations at 0 dB SNR with 300 epochs

The confusion matrix performance of the proposed deep learning mode for digital and analog modulation signals at 12 dB SNR with 300 epochs is shown in Fig. 4.19.

The confusion matrix performance of the proposed deep learning mode for digital modulation signals at 14 dB SNR with 50 epochs is shown in Fig. 4.20.

The confusion matrix performance of the proposed deep learning model for digital and analog modulation signals at 14 dB SNR with 300 epochs is shown in Fig. 4.21.

The confusion matrix performance of the proposed deep learning mode for digital modulation signals at 16 dB SNR with 50 epochs is shown in Fig. 4.22.

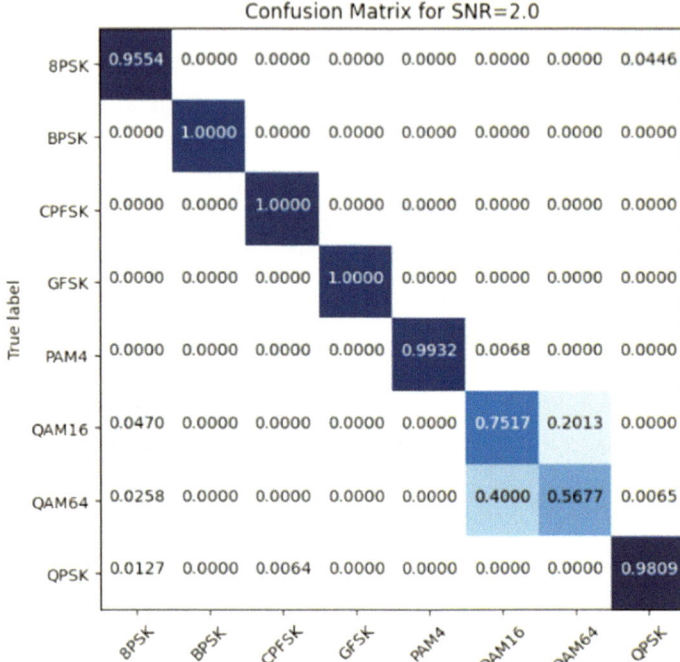

Fig. 4.6 Confusion Matrix of our proposed deep learning model for digital modulations at 2 dB SNR with 50 epochs

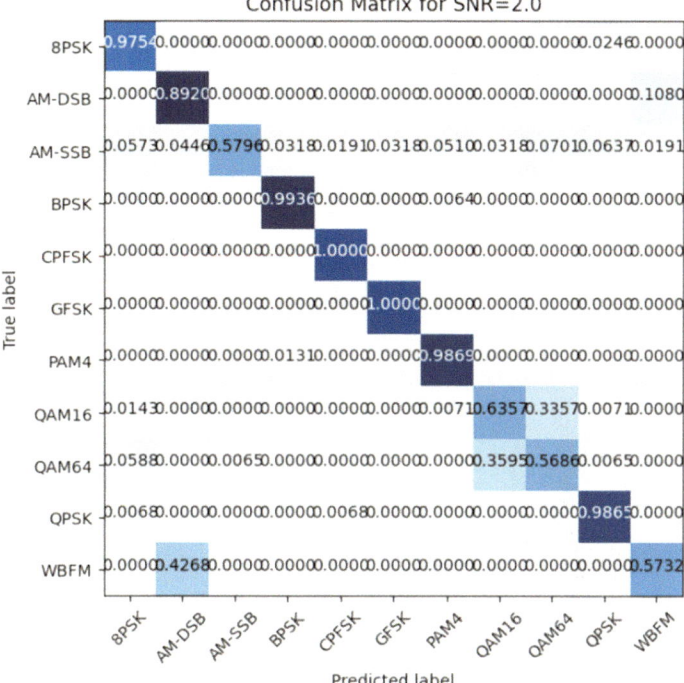

Fig. 4.7 Confusion Matrix of our proposed deep learning model for both digital and analog modulations at 2 dB SNR with 300 epochs

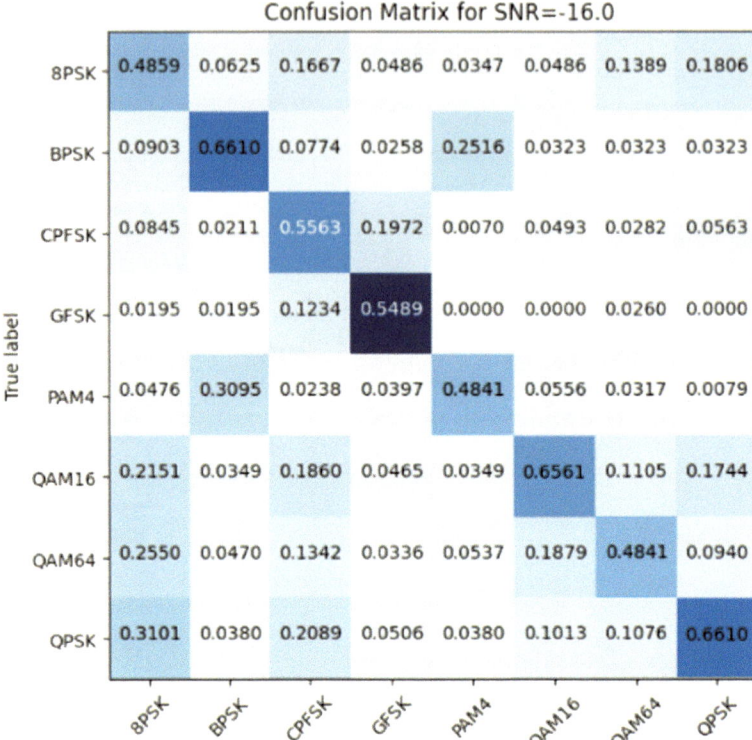

Fig. 4.8 Confusion Matrix of our proposed deep learning model for digital modulations at $-16\,dB$ SNR with 50 epochs

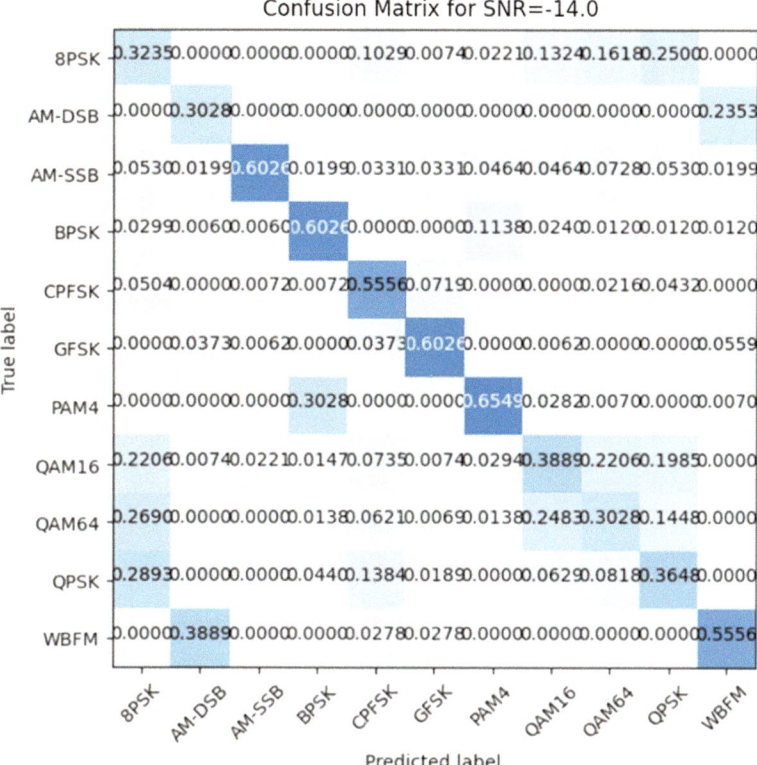

Fig. 4.9 Confusion Matrix of our proposed deep learning model for both digital and analog modulations at -14 dB SNR with 300 epochs

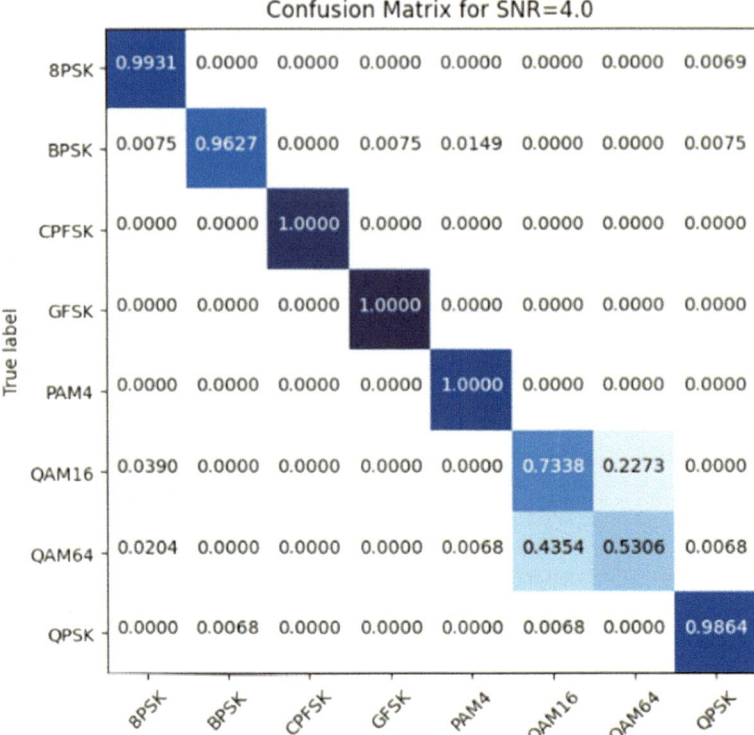

Fig. 4.10 Confusion Matrix of our proposed deep learning model for digital modulations at 4 dB SNR with 50 epochs

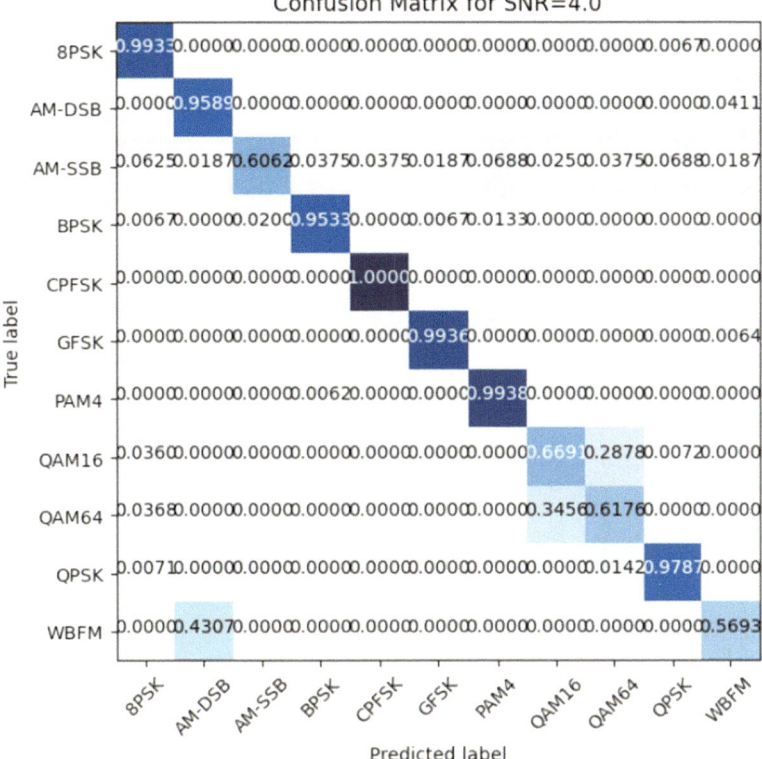

Fig. 4.11 Confusion Matrix of our proposed deep learning model for both digital and analog modulations at 4 dB SNR with 300 epochs

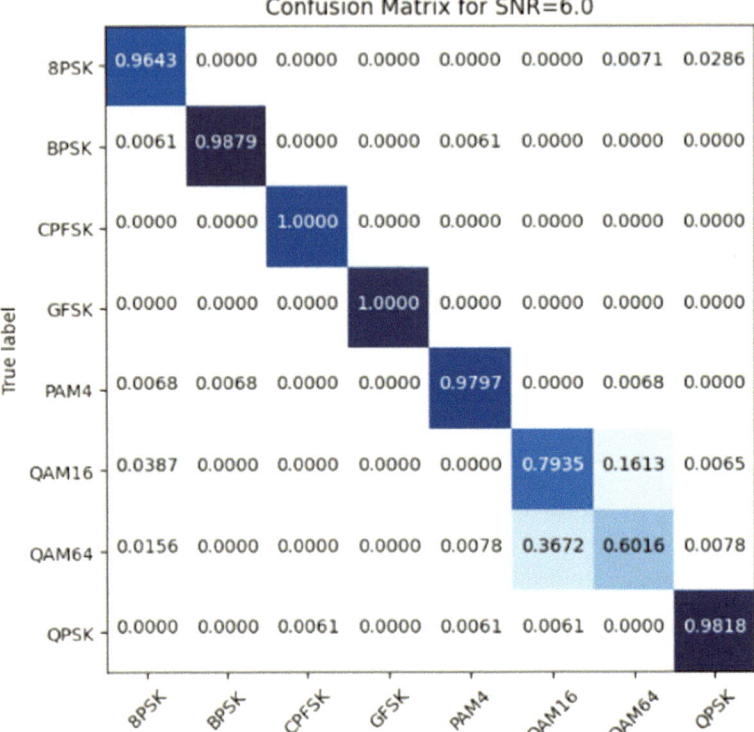

Fig. 4.12 Confusion Matrix of our proposed deep learning model for digital modulations at 6 dB SNR with 50 epochs

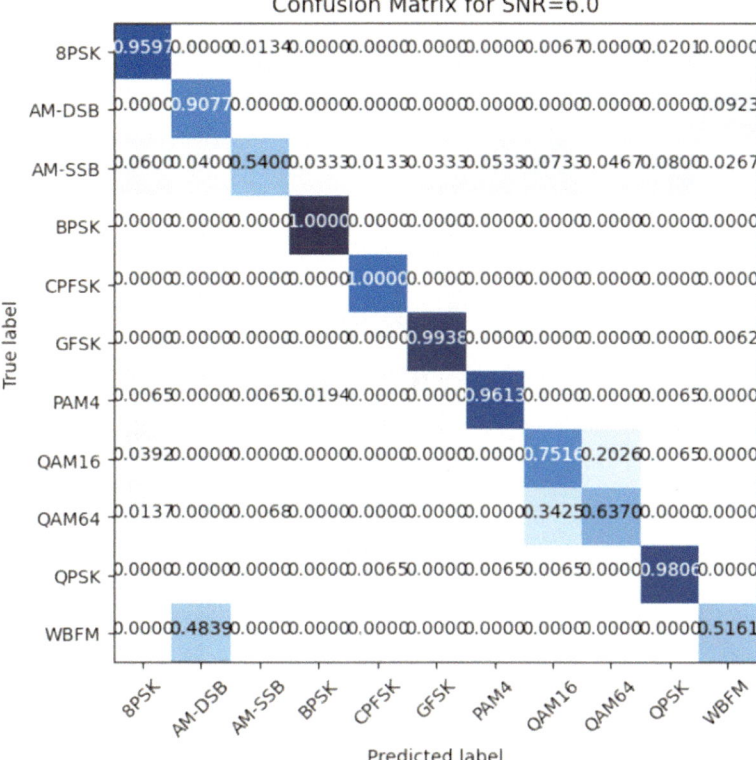

Fig. 4.13 Confusion Matrix of our proposed deep learning model for both digital and analog modulations at 6 dB SNR with 300 epochs

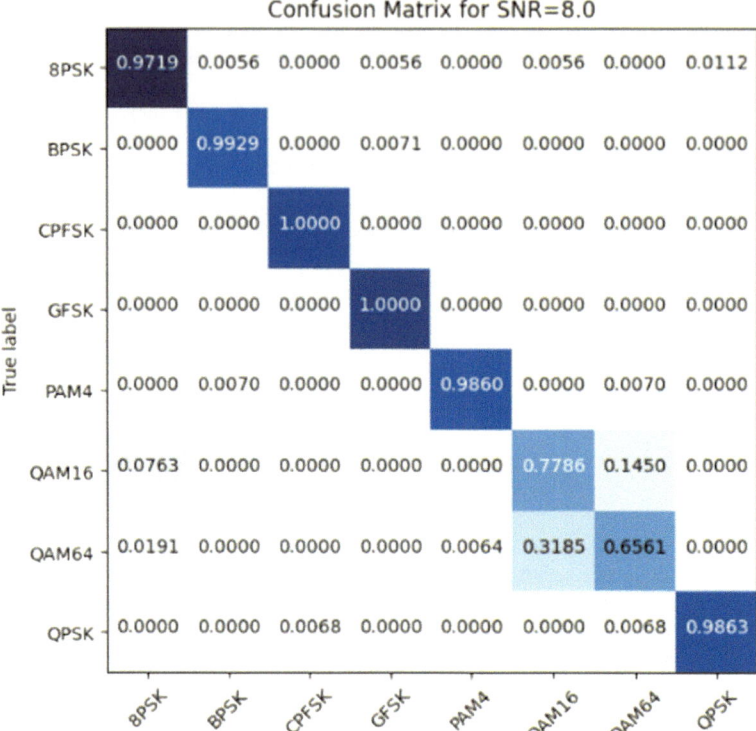

Fig. 4.14 Confusion Matrix of our proposed deep learning model for digital modulations at 8 dB SNR with 50 epochs

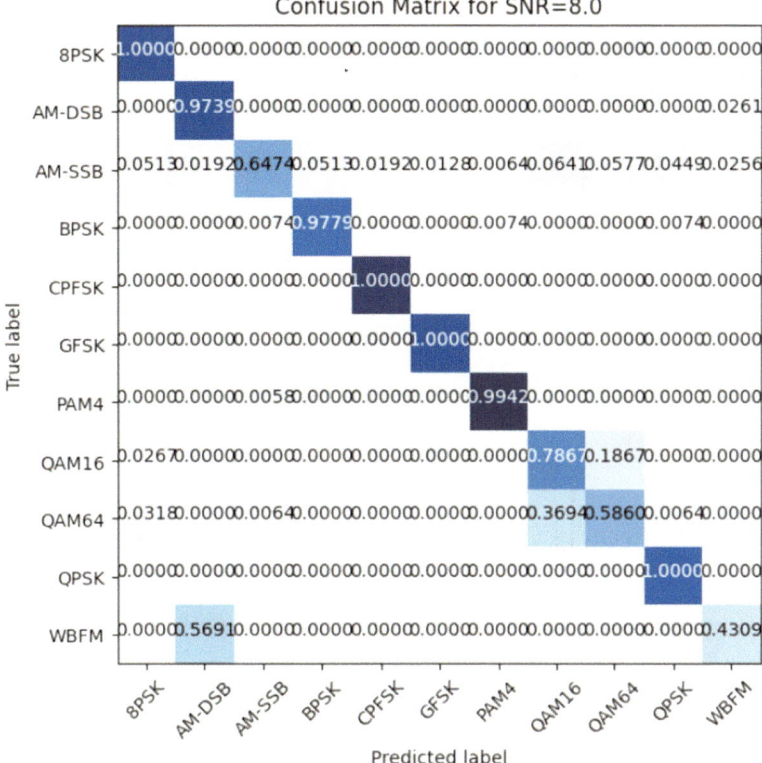

Fig. 4.15 Confusion Matrix of our proposed deep learning model for both digital and analog modulations at 8 dB SNR with 300 epochs

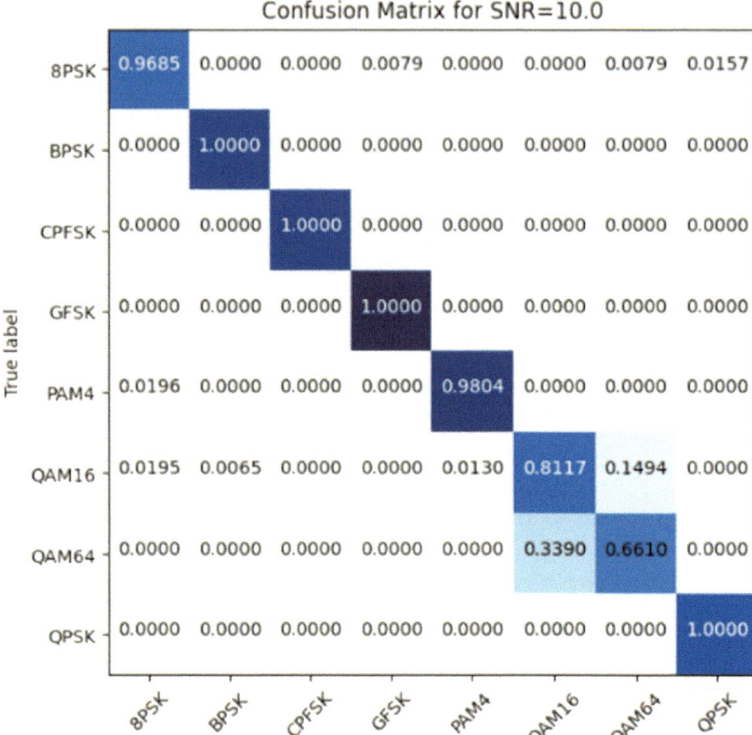

Fig. 4.16 Confusion Matrix of our proposed deep learning model for digital modulations at 10 dB SNR with 50 epochs

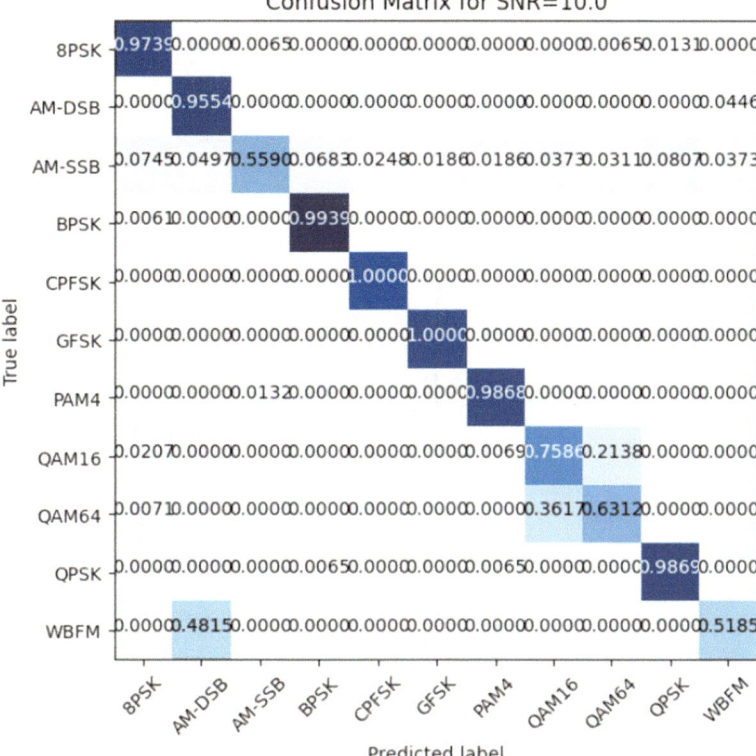

Fig. 4.17 Confusion Matrix of our proposed deep learning model for both digital and analog modulations at 10 dB SNR with 300 epochs

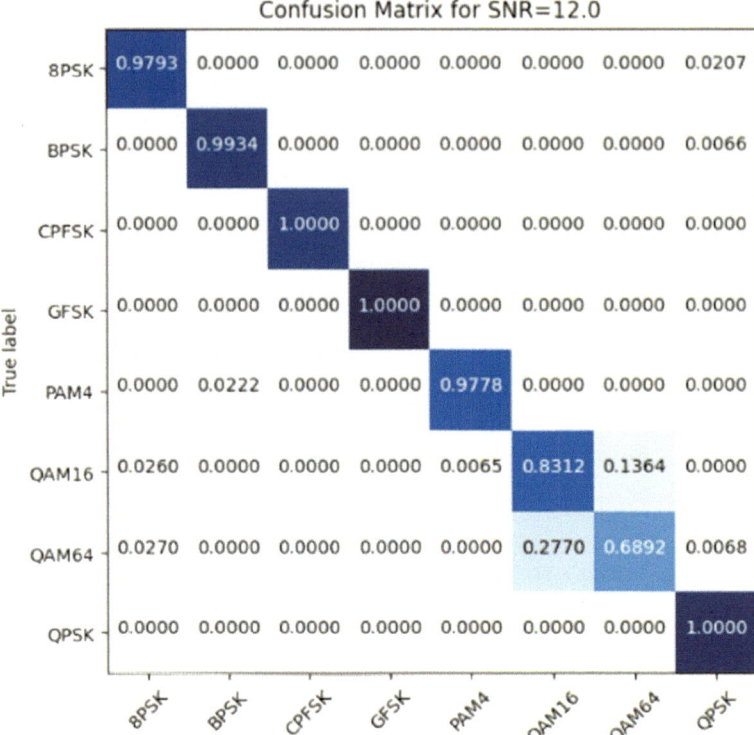

Fig. 4.18 Confusion Matrix of our proposed deep learning model for digital modulations at 12 dB SNR with 50 epochs

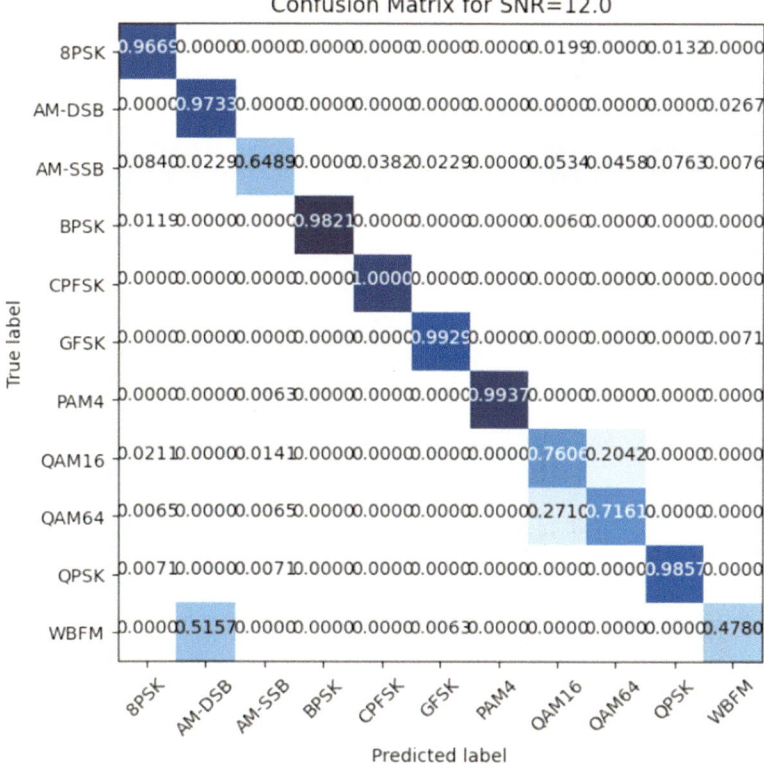

Fig. 4.19 Confusion Matrix of our proposed deep learning model for both digital and analog modulations at 12 dB SNR with 300 epochs

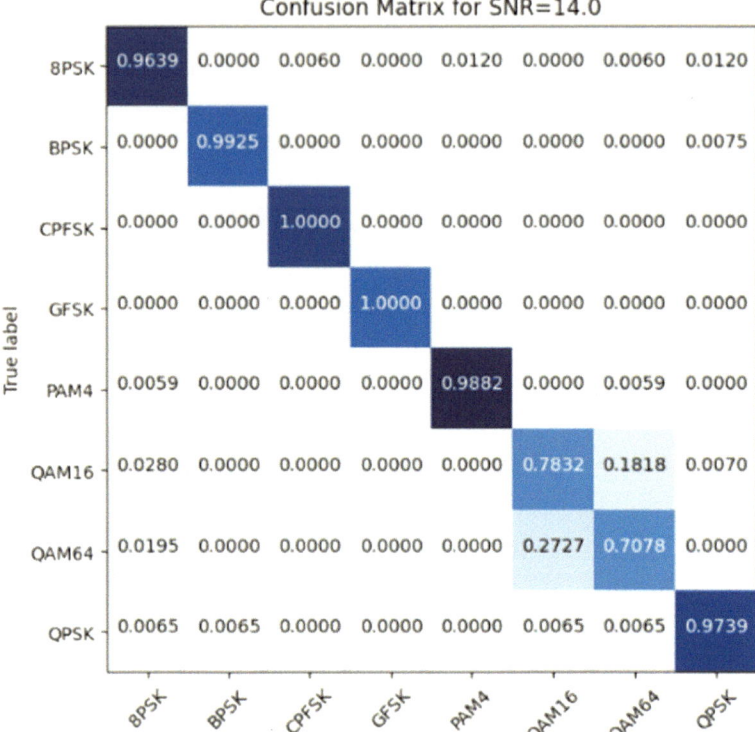

Fig. 4.20 Confusion Matrix of our proposed deep learning model for digital modulations at 14 dB SNR with 50 epochs

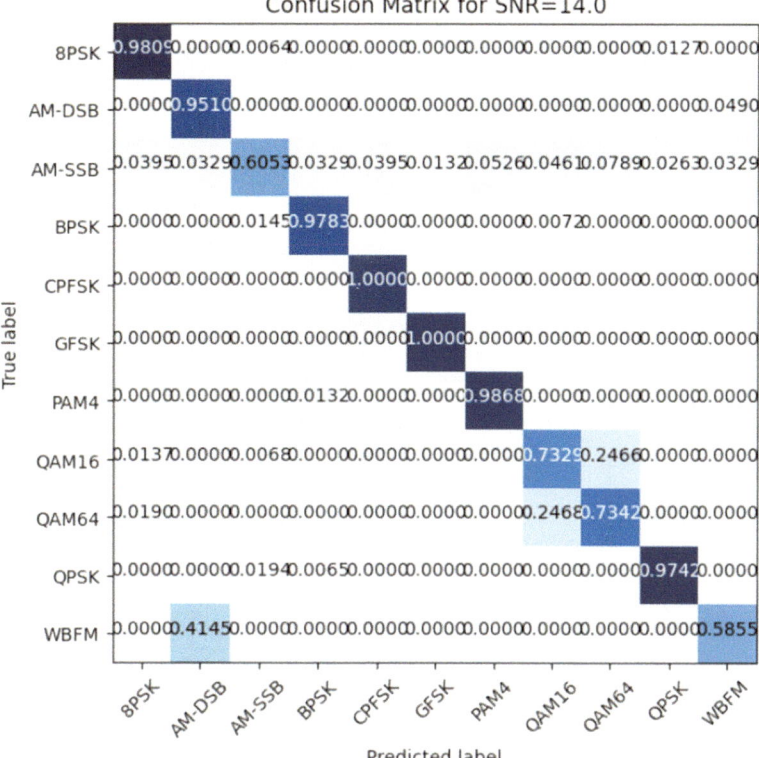

Fig. 4.21 Confusion Matrix of our proposed deep learning model for both digital and analog modulations at 14 dB SNR with 300 epochs

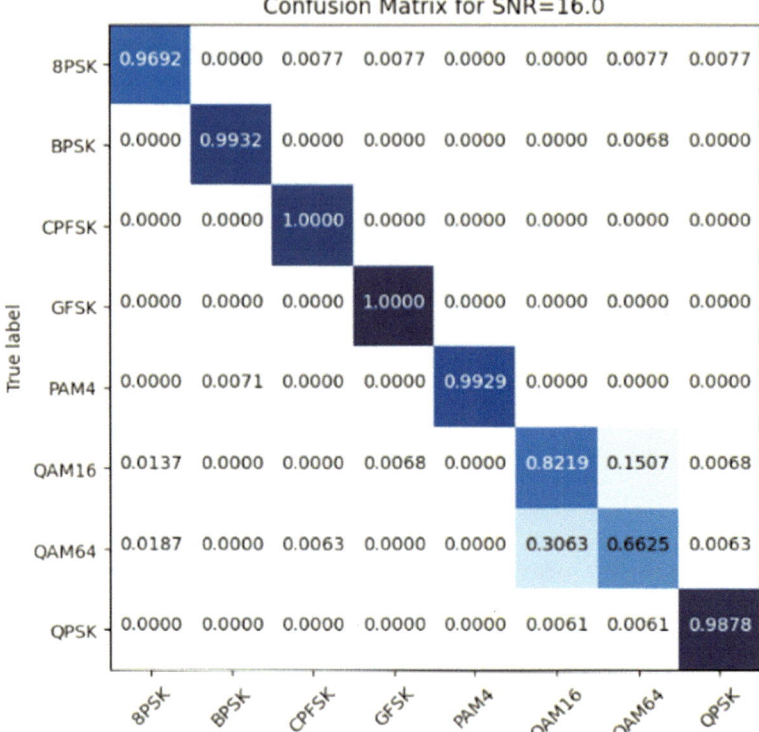

Fig. 4.22 Confusion Matrix of our proposed deep learning model for digital modulations at 16 dB SNR with 50 epochs

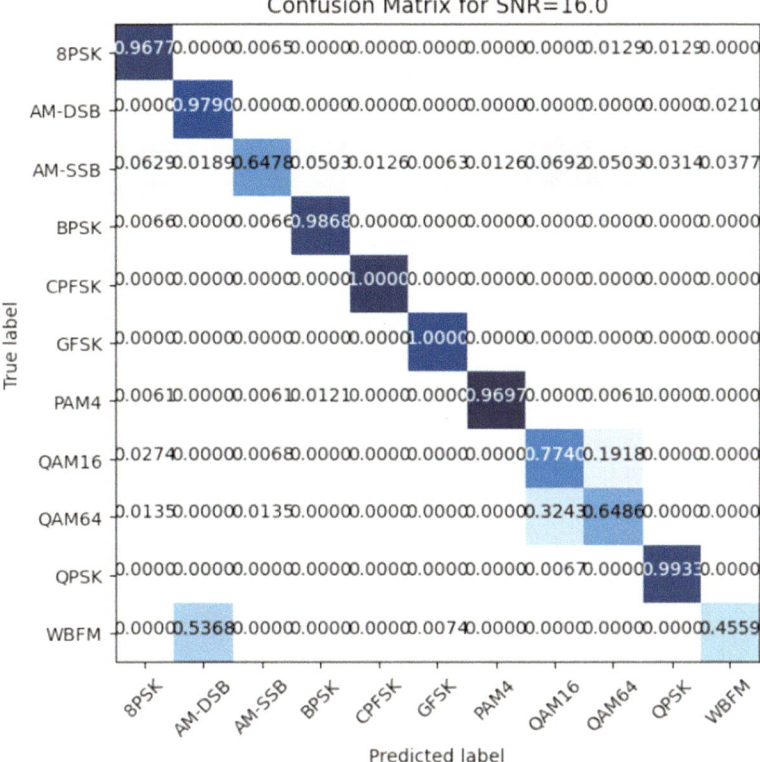

Fig. 4.23 Confusion Matrix of our proposed deep learning model for digital and analog modulations at 16 dB SNR with 300 epochs

The confusion matrix performance of the proposed deep learning mode for digital and analog modulation signals at 16 dB SNR with 300 epochs is shown in Fig. 4.23.

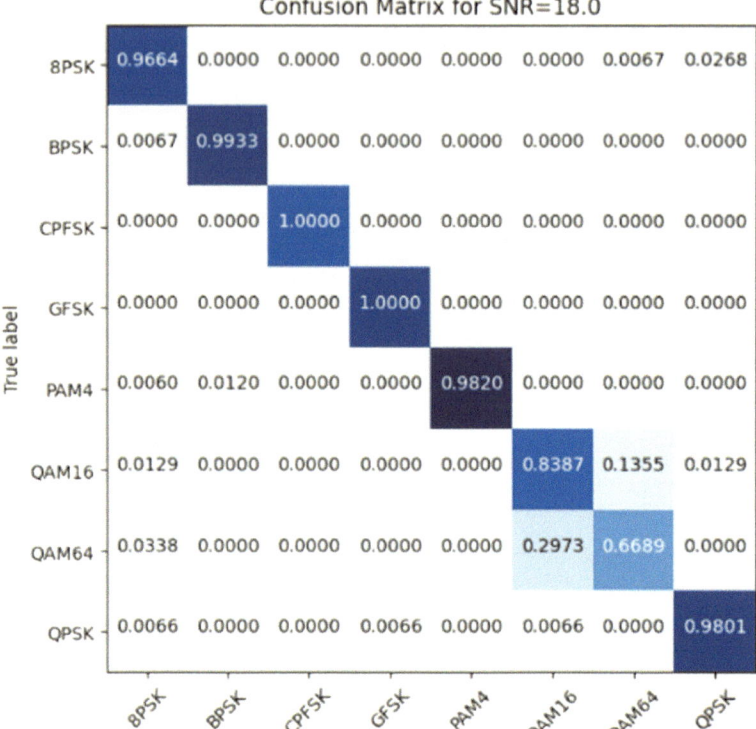

Fig. 4.24 Confusion Matrix of our proposed deep learning model for digital modulations at 18 dB SNR with 50 epochs

The confusion matrix performance of the proposed deep learning mode for signals at 18 dB SNR with 50 epochs is shown in Fig. 4.24.

The confusion matrix performance of the proposed deep learning mode for digital and analog modulation signals at 18 dB SNR with 300 epochs is shown in Fig. 4.25.

The results of the confusion matrix in Figs. 4.4 to 4.25 show that our proposed model performs well across most signal types as shown in the confusion matrix in Figs. 4.4 to 4.25. We analyze the classification accuracy of our proposed model for different modulation with DeepSig RadioML2016.10a dataset.

It is noted that QAM16 is often misrecognized as QAM64 and vice versa because the constellation points of QAM16 can be found in the constellation points of QAM64 so they can have constellation in common which causes short time observation to suffer. Moreover, features for a signal with QAM64 modulation may not be captured by just 128 samples and so the deep network confuses it with QAM16 and therefore might benefit from a bigger dataset.

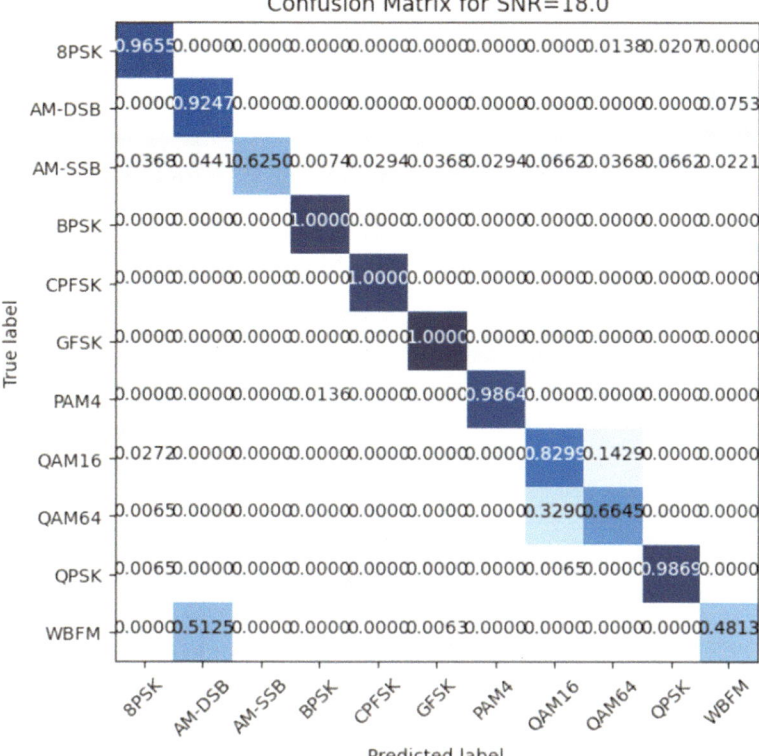

Fig. 4.25 Confusion Matrix of our proposed deep learning model for digital and analog modulations at 18 dB SNR with 300 epochs

The accuracy performance of the proposed deep learning model per various modulation type versus SNR is shown in Fig. 4.26. A noisy signal will have a low SNR that means that if the noise is higher, the model will likely to less accuracy to do the modulation classification.

The overall classification performance of the proposed combined deep learning model is shown in Table 4.1.

The accuracy performance of the proposed deep learning model per various modulation type versus SNR from −20 to −2 dB is shown in Table 4.2.

The accuracy performance of the proposed deep learning model per various modulation type versus SNR from 0 to 20 dB is shown in Table 4.3.

The overall training and prediction time performance of the proposed combined deep learning model compared to other models is shown in Table 4.4.

Other published work [25] that do not use deep learning models do not have the capability of applying automatic modulation recognition prediction. Other [42] published work that use deep learning models only achieve high accuracy in high SNR signals only unlike our

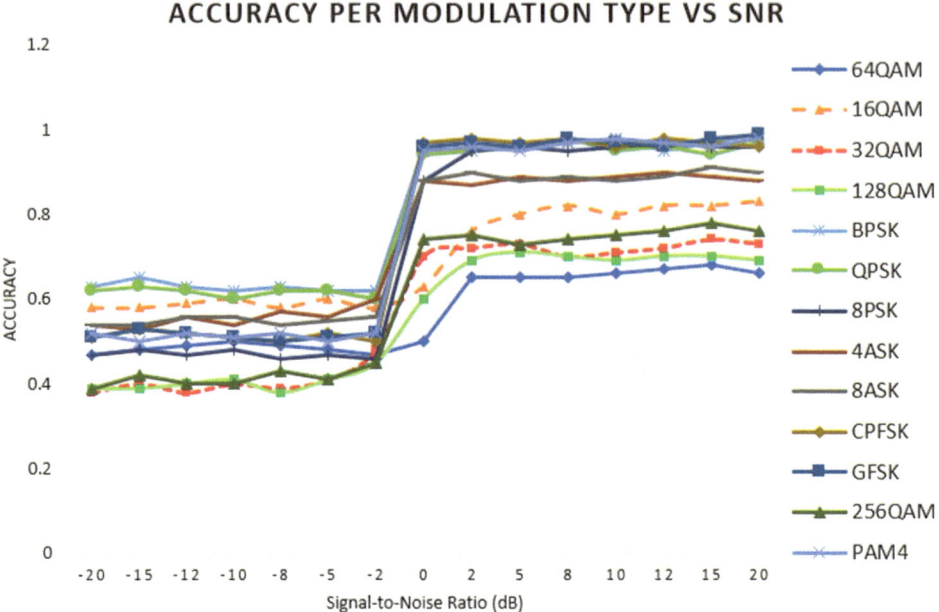

Fig. 4.26 Accuracy performance of the proposed deep learning model per various modulation type versus SNR

Table 4.1 Overall classification performance of the proposed combinatorial deep learning model

SNR	Model	Accuracy
High SNR signals	ConvLSTM	93.5% for High SNR signals
Low SNR signals	Transformer-block	62% for Low SNR signals
Both high and	Transformer-block + ConvLSTM	62% for Low SNR and 93.5% for High SNR Signals

proposed work where we achieve high accuracy in both high and low SNR signals. One of the main advantages of our proposed work is that our combined deep learning model provide the capability of loading automatic modulation recognition on hardware accelerator chips to take processing load of the main hardware processor. Also, another advantage of our proposed work is that by utilizing Transformer-block processing utilized for larger training data set parallelization in our combined model resulting in faster training time and inference testing as shown in Table 4.5. However, does not achieve the fastest prediction and training time compared combined models which is one of the shortcomings of our combined deep learning model (Fig. 4.27).

As can be seen in Table 4.5 Hao [18] used CLDNN+GRU model achieving accuracy of 90% at 0 dB SNR and less than 20% at −16 dB SNR. Huang [4] used Comprehensive CNN

Table 4.2 Accuracy performance per various modulation type versus SNR

Modulation type	−20	−15	−12	−10	−8	−5	−2
16QAM	0.58	0.58	0.59	0.60	0.58	0.60	0.58
32QAM	0.40	0.41	0.39	0.40	0.39	0.41	0.48
64QAM	0.47	0.47	0.49	0.50	0.49	0.48	0.47
128QAM	0.41	0.40	0.40	0.41	0.38	0.41	0.45
256QAM	0.41	0.42	0.40	0.40	0.43	0.41	0.45
BPSK	0.63	0.65	0.63	0.62	0.63	0.62	0.62
QPSK	0.62	0.63	0.62	0.60	0.62	0.62	0.60
8PSK	0.47	0.48	0.47	0.48	0.46	0.47	0.46
4ASK	0.54	0.53	0.56	0.54	0.57	0.54	0.60
8ASK	0.54	0.54	0.56	0.56	0.54	0.55	0.56
PAM4	0.52	0.51	0.52	0.51	0.52	0.50	0.52
CPFSK	0.51	0.53	0.52	0.51	0.50	0.52	0.50
GFSK	0.51	0.53	0.52	0.51	0.50	0.51	0.52

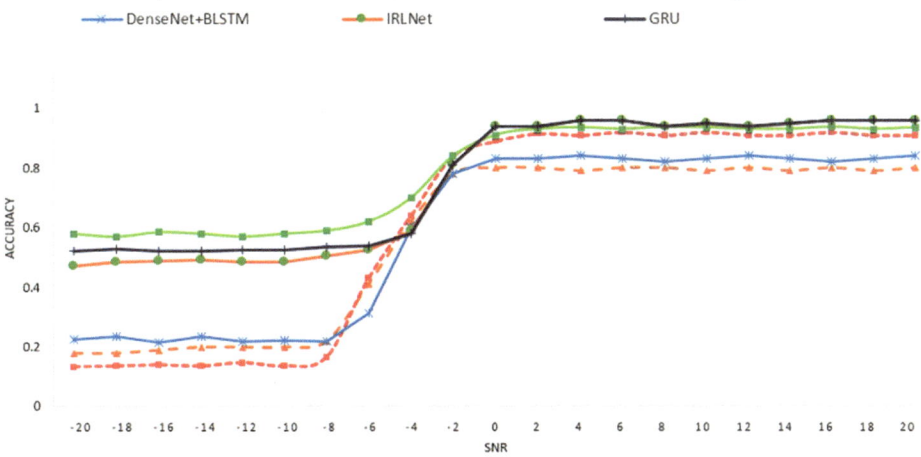

Fig.4.27 Accuracy performance of the proposed deep learning model per various Models type versus SNR

model achieving accuracy of 80% at 0 dB SNR and less than 20% at −16 dB SNR. Xie [45] used DenseNet and BLSTM model achieving accuracy of 84% at 0 dB SNR and less than 25% at −16 dB SNR. Yang [55] used IRLNet model achieving accuracy of 97% at 5 dB SNR

Table 4.3 Accuracy performance per various modulation type versus SNR

Modulation type	0	2	5	8	10	15	20
16QAM	0.63	0.76	0.80	0.82	0.66	0.82	0.83
32QAM	0.70	0.72	0.73	0.70	0.71	0.74	0.73
64QAM	0.50	0.65	0.65	0.65	0.66	0.68	0.66
128QAM	0.60	0.69	0.71	0.70	0.69	0.70	0.70
256QAM	0.74	0.75	0.73	0.74	0.75	0.78	0.76
BPSK	0.96	0.95	0.96	0.97	0.98	0.97	0.99
QPSK	0.94	0.95	0.96	0.98	0.95	0.94	0.97
8PSK	0.88	0.95	0.96	0.95	0.96	0.96	0.96
4ASK	0.88	0.87	0.89	0.88	0.89	0.89	0.88
8ASK	0.88	0.90	0.88	0.89	0.88	0.91	0.90
PAM4	0.95	0.96	0.95	0.97	0.98	0.96	0.98
CPFSK	0.97	0.98	0.97	0.98	0.96	0.97	0.96
GFSK	0.96	0.97	0.96	0.98	0.97	0.98	0.99

Table 4.4 Overall training & prediction time performance

Model	Training time per epoch (s)	Prediction time (μ/Sample)	Training time (s)
CNN [45]	30	1000	600
LSTM [45]	800	2000	44800
SCRNN [45]	280	600	11480
ResNet [54]	341	183	1850
Transformer-block + ConvLSTM (this work)	320	730	2250

and less than 50% at -16 dB SNR. Liu [42] used GRU model achieving accuracy of 86% at 0 dB SNR and less than 55% at -16 dB SNR. Our proposed model used ConvLSTM and Transformer-block model achieving accuracy of 93.5% at 0 dB SNR and 62% at -16 dB SNR.

As shown in Table 4.5, our proposed model used ConvLSTM and Transformer-block model achieving accuracy of 93.5% at 0 dB SNR and 62% at -16 dB SNR. Whereas Liu [60] used GRU model achieving accuracy of 86% at 0 dB SNR and less than 55% at -16 dB SNR. Yang [61] used IRLNet model achieving accuracy of 97% at 5 dB SNR and less than 50% at -16 dB SNR. Xie [62] used DenseNet and BLSTM model achieving accuracy of 84% at 0 dB SNR and less than 25% at -16 dB SNR. And Huang [63] used Comprehensive

Table 4.5 Overall classification modulation recognition accuracy performance comparison

Author	Input signal	Model	Recognition accuracy
Jiang [12]	IQ sequence	CNN + Bi LSTM	90%(0 dB) & <10%(−16 dB)
Tang [10]	constellation map	CNN + GAN	100%(−2 dB)& <10%(−16 dB)
Xu [11]	IQ sequence	CNN + LSTM+ FC	90%(0 dB) & <10%(−16 dB)
Liang [13]	IQ sequence	ResNeXt + Attention	90%(0 dB) & <10%(−16 dB)
Zhang [15]	Sampled signal	GRU + CNN	99.45%(0 dB) & <10%(−16 dB)
Chang [14]	Sampled signal	CNN + Bi GRU + SAFN	84%(0 dB) & <10%(−16 dB)
Liu [20]	IQ sequence	DCN + BiLSTM	90%(0 dB) & <10%(−16 dB)
Bai [17]	IQ sequence	DMFF + CNN	90%(0 dB) & <10%(−16 dB)
Udaya [19]	IQ sequence	LSTM + Bi LSTM	90%(0 dB) & <10%(−16 dB)
Zou [16]	IQ sequence	Attention + CLDNN	90%(4 dB) & <10%(−16 dB)
Hao [18]	IQ sequence	CLDNN+GRU	90%(0 dB) & <20%(−16 dB)
Huang [4]	IQ sequence	Compressive CNN	80%(0 dB) & <20%(−16 dB)
Xie [45]	IQ sequence	DenseNet-BLSTM	84%(0 dB) & <25%(−16 dB)
Yang [55]	IQ sequence	IRLNET	97%(5 dB) & <50%(−16 dB)
Liu [42]	IQ sequence	GRU	86%(0 dB) & <55%(−16 dB)
(This work)	IQ sequence	Transformer-block + ConvLSTM	93.5%(0 dB) & 62%(−16 dB)

CNN model achieving accuracy of 80% at 0 dB SNR and less than 20% at −16 dB SNR. And Hao [64] used CLDNN+GRU model achieving accuracy of 90% at 0 dB SNR and less than 20% at −16 dB SNR. All other published models in Table 4.5 achieve less than 10% at −16 dB SNR compared to our proposed model.

Table 4.6 Model layer number of parameters

Model layer	Trainable parameter	Number of flops	Memory (MB)
CNN [23]	5456219	80548043	61.4
LSTM [23]	199563	7696283	2.31
This work	1702625	60662829	19.7

We compare our proposed models in terms of the number of trainable parameters, the number of floating point operations (FLOPs) and the memory cost as shown in Table 4.6. Smaller number of trainable parameters requires fewer FLOPs and the smaller memory space.

It can be seen in Table 4.6 that our proposed model needs to train 1702625 parameters with total number of 60662829 floating point operations (FLOPs) at a 19.7MB memory cost. Whereas LTSM model in [65] needs to train 199563 parameters with total number of 7696283 floating point operations (FLOPs) at a 2.31MB memory cost. Also, CNN [65] needs to train 5456219 parameters with total number of 80548043 floating point operations (FLOPs) at a 61.4 MB memory cost as shown in Table 4.6.

Spiking Neural Networks Neuromorphic Computational System

5

5.1 Spiking Neural Networks Neuromorphic Computational System Modeling

Neuromorphic computing has the potential to be the implementation of choice for low-power Deep Learning systems. Spiking neural networks are regarded as the third generation of artificial neural networks (ANNs). SNNs are more biologically plausible and can be more energy-efficient, especially for applications in neuromorphic computing. Generally, SNNs tend to have lower accuracy compared to their ANN counterparts, but they offer advantages in computational efficiency and energy consumption. ANN-to-SNN refers to the process of converting an Artificial Neural Network (ANN) to a Spiking Neural Network (SNN). ANN-to-SNN conversions allows for the efficient and effective training of SNNs by leveraging the knowledge captured in pre-trained ANNs [32–34].

The transformation process from ANN to SNN as shown in Fig. 5.1 starts with ANN is first trained using standard techniques such as backpropagation to learn weights and biases. Then the learned weights are then adjusted or quantized to match the requirements of the SNN framework. During inference, the SNN processes inputs using spikes, and the output is derived based on the spiking activity of the neurons. ANNs operate with continuous values and typically use gradient-based learning rules such as backpropagation. In contrast, SNNs use spikes as their primary means of communication, where neurons fire discrete events based on certain thresholds. The process of converting an ANN to an SNN, which involves adapting learning rules and the representation of information from continuous values to discrete spikes. This transformation involves adapting the continuous activation functions and learning rules of ANNs into the discrete event-driven framework of SNNs [32–34].

In the context of converting an Artificial Neural Network (ANN) to a Spiking Neural Network (SNN), transfer learning plays a crucial role in leveraging the knowledge gained from training the ANN to improve the performance and efficiency of the SNN. Transfer

Z. El-Khatib and S. Moussa, *Wireless Communication Using Deep Learning Techniques for Neuromorphic VLSI Computing*, Synthesis Lectures on Engineering, Science, and Technology, https://doi.org/10.1007/978-3-031-73800-5_5

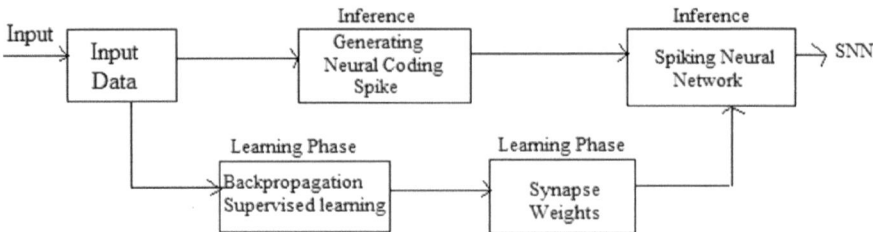

Fig. 5.1 Artificial Neural Networks (ANN) to Spiking Neural Networks (SNN) conversion

Learning is a technique in machine learning where a model developed for one task is reused as the starting point for a model on a second task. In transfer learning, a pre-trained ANN is used. This model has already been trained on a large dataset, allowing it to capture useful features and representations. The learned weights and biases can be transferred to the SNN, allowing it to start with a good initialization. After transferring the weights, the SNN can be further trained on specific tasks, often using methods like spike-timing dependent plasticity (STDP) or other learning rules designed for spiking networks. First convert the real-valued weights from the ANN to synaptic weights for the SNN, this often involves quantization. Then use a suitable encoding scheme to convert the input data into spike trains. Using STDP to adjust synaptic weights based on the timing of spikes. The key steps involve training the ANN, defining the LIF neuron model, transferring weights, encoding inputs into spike formats, fine-tuning the SNN, then using a pre-trained ANN can significantly reduce the time and data required to train the SNN [32–35].

The learning rules for transforming an ANN to an SNN often involve Spike-Timing Dependent Plasticity (STDP) A biological-inspired learning rule where the strength of connections between neurons is adjusted based on the relative timing of spikes. The weight adaption process where the weights from the ANN are often quantized or adjusted to fit the requirements of the SNN framework. This might involve modifying the continuous weights of the ANN into values suitable for the spiking nature of SNNs. After transferring the weights, the SNN can be further trained on specific tasks, often using methods like spike-timing dependent plasticity (STDP) or other learning rules designed for spiking networks. Table 5.1 shows different ANN-to-SNN Learning Models Accuracy Performance comparison. Using a pre-trained ANN can significantly reduce the time and data required to train the SNN and can be more efficient [32–35].

Spiking Neural Networks (SNN) is used to connect machine learning and neuroscience. Spiking neural networks (SNN) provides a promising solution for low-power hardware for neuromorphic computing. Using Spiking Neural Network circuit methods and doing parallel computations can reduce costs. Spiking neural network is more promising than other neural networks that can pave a new way for low-power neuromorphic computing applications. The brain's energy efficiency for decision making cognitive tasks made scientists to focus their efforts on building non-Von Neumann computer systems that imitate the biological brain.

Table 5.1 ANN-to-SNN learning models accuracy performance comparison

Neuron type	Model type	Network architecture	Recognition accuracy (SSN) (%)
LIF [34]	CNN-to-SNN	Convolutional Neural Network	96
LIF [35]	ANN-to-SNN	3Convn 2Linear	77
LIF [32]	ANN-to-SNN	VGG16	91.5
LIF [33]	ANN-to-SNN	VGG9	90.5

Neurons process information as asynchronous event-driven spikes and retain memories as synaptic strengths of their connection in the brain. Because transistors have properties similar to nerve membrane channels. When transistors are operated in weak inversion region, they leak a very small current. This transistor region of operation is also known as the subthreshold region. This way a large network of thousands of neurons will consume very low power. Spiking Neural Networks circuit functions with a pre-trained network's weights consume less power [65–71].

Biological neurons are in a network where a neuron receives input from another neuron. The inputs are received as a spike in the synapse. The in turn induce an output at the post synaptic. A Spiking Neural Network system consist of the following circuit building blocks as shown in Fig. 5.2. The input receives the analog vector input signals [72, 73]. The weighted multiplication multiply the input vector elements by corresponding weights in an analog vector matrix multiplication (VMM). The analog vector matrix multiplication input is fed to a Winner-Take-All (WTA) circuit. The output from the Winner-Take-All (WTA) circuit is connected to a Differential-Pair Integrator (DPI) synapse circuit [74, 75]. The Differential-Pair Integrator (DPI) synapse circuit is then connected to an Integrate-and-Fire neuron. The output from the Integrate-and-Fire neuron is connected to the spike-timing dependent plasticity (STDP) circuit as shown in Fig. 5.2. The DPI synapse integrates the incoming signals [65–68]. The WTA circuit selects the most active signal among neurons. The Integrate-and-Fire neuron integrates signals over time and generates output spikes. Then the STDP circuit adjusts synaptic weights based on the timing of spiking activity for learning and plasticity [69–71, 73–75].

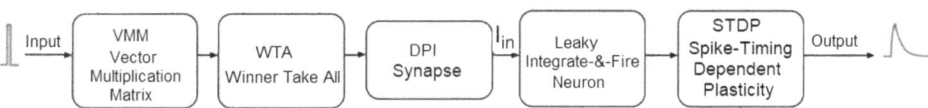

Fig. 5.2 Spiking Neural Network system

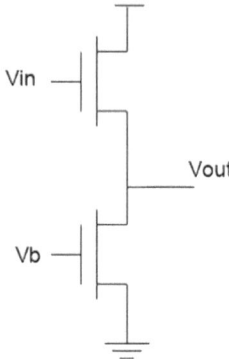

Fig. 5.3 Source follower circuit for SNN

Analog VLSI is utilized to design spiking neural networks circuits such as silicon synapse and CMOS neuron. Because transistors have properties similar to nerve membrane channels. When transistors are operated in weak inversion region they leak a very small current. This transistor region of operation is also known as the subthreshold region. This way a large network of thousands of neurons will consume very low power. Spiking Neural Networks (SNN) do not fire continuously. SNN fires only when the post-synaptic potential reaches a certain threshold value. In the spiking neural networks circuits the transistors are operating at the subthreshold level (weak inversion region). At this mode of operation, the current-voltage relationship is exponential and is best described by an exponential drain current equation.

Equations of subthreshold nfet transistor for a source follower circuit for SNN shown in Fig. 5.3 can be described as follows [30, 76, 77]

$$I = I_0 e^{\kappa V_g / U_T} \left(e^{-V_s / U_T} - e^{-V_d / U_T} \right) \tag{5.1.1}$$

where κ is given by the following Eq. 5.1.2 [30, 76, 77]

$$\kappa (\text{kappa}) = \frac{\partial \psi_s}{\partial V_g} = \frac{C_{\text{ox}}}{C_{\text{ox}} + C_{\text{dep}}} \tag{5.1.2}$$

Equations of subthreshold pfet transistor for a source follower circuit for SNN shown in Fig. 5.3 can be described as follows [30, 76, 77]

$$I = I_0 e^{-\kappa V_g / U_T} \left(e^{V_s / U_T} - e^{V_d / U_T} \right) \tag{5.1.3}$$

Synapses are responsible for connecting neurons and communicating spike signals between them. A synapse receives spike voltages from the output of its pre-synaptic neuron. It produces a current based on a weight value [65–69]. Then it feeds this weighted current to its post-synaptic neuron [32, 70, 71, 73–75, 78, 79].

Exponentially decaying Log-domain Pulse Integrator is shown in Fig. 5.4. The linear integrator response is of a low-pass filter with a decaying exponential [65, 67, 80–82].

Fig. 5.4 Log-domain pulse
integrator circuit

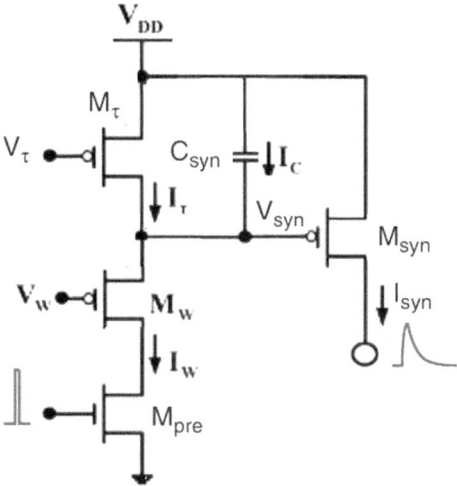

By changing the bias voltage V_W the synaptic weight can be varied. The current $Isyn$ can be varied up and down exponentially with time [65, 67, 80–82].

$$I_{syn}(t) = \begin{cases} I_{syn}^- e^{+\frac{(t-t_i^-)}{\tau_c}} & \text{(charge phase)} \\ I_{syn}^+ e^{-\frac{(t-t_i^+)}{\tau_d}} & \text{(discharge phase)} \end{cases}$$ (5.1.4)

$$I_{syn}(t) = I_0 e^{-\frac{\tau_c - f\Delta t(\tau_c + \tau_d)}{\tau_c \tau_d}t}$$

The δ_t is the time duration and the n is the number of pulses. The following Eqs. 5.1.4 and 5.1.5 describe the currents in sub-threshold mode of operation [65, 67, 73–75, 83].

$$I_w = I_0 e^{\frac{k}{U_T}(V_{syn} - V_w)}$$

$$I_\tau = I_0 e^{\frac{\kappa(V_{dd} - V_\tau)}{U_T}}$$

$$I_c = C\frac{d}{dt}(V_{dd} - V_{syn})$$ (5.1.5)

$$I_{syn} = I_0 e^{\frac{\kappa(V_{dd} - V_{syn})}{U_T}}$$

The derivative of the current through a capacitor can be determined by the following Eq. 5.1.6. The larger the time-constant τ the larger the capacitor as shown in Eq. 5.1.5 [65, 67, 73–75, 83].

The solution for the first-order differential Equation 5.1.6 is given by Eq. 5.1.6. For a larger synapse weight we have to have a large I_W current [65, 67, 73–75, 83].

$$\frac{d}{dt}I_{syn} = -I_{syn}\frac{\kappa}{U_T}\frac{d}{dt}V_{syn}$$

$$\tau\frac{d}{dt}I_{\text{syn}} = -I_{\text{syn}} + I_{\text{syn}}\frac{I_w}{I_\tau}, \quad \tau = \frac{C_{\text{syn}}U_T}{\kappa I_\tau}$$

$$I_w = I_0 e^{-\frac{\kappa(V_w - V_{\text{syn}})}{U_T}}$$

$$= I_0 e^{-\frac{k(v_w - V_{dd})}{U_T}} e^{\frac{k(V_{ssn} - V_{dd})}{U_T}}$$

$$= I_{w0}\frac{I_0}{I_{\text{syn}}}$$

(5.1.6)

To get a larger synapse weight we can implement a differential-pair integrator (DPI) synapse circuit. The differential-pair integrator (DPI) synapse circuit is shown in Fig. 5.5 [65, 67, 73–75].

$$I_{\text{out}} = I_0 e^{\frac{kV_C}{U_T}}$$

$$I_1 + I_2 = I_{\text{in}}$$

$$I_2 = I_\tau + I_C$$

$$I_C = C\frac{d}{dt}V_C$$

$$I_C = C\frac{U_T}{\kappa I_{\text{out}}}\frac{d}{dt}I_{\text{out}}$$

$$\tau = \frac{CU_T}{\kappa I_\tau}$$

(5.1.7)

Fig. 5.5 Differential pair integrator synapse circuit

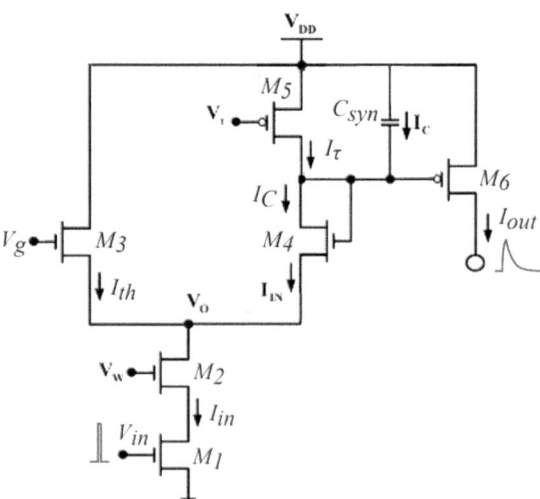

The differential-pair integrator (DPI) synapse integrates incoming signals as shown in Fig. 5.2. It compares two input signals and generates an output based on their difference [65, 67, 73–75].

We can implement a Differential-Pair integrator synapse by connecting the bulk of the pmos transistor M_W to its source hence isolating its well. The capacitor C voltage $Vsyn$ is shown in the circuit in Fig. 5.5.

The weighted contribution of a synapse can be implemented using a pmos transistor for excitator synapse and nmos for inhibitory synapse. In the differential-pair integrator (DPI) synapse, Vg controls the gain of the integration [65, 67, 73–75].

The current Iin is based on the assigned weight V_w. Then $Csyn$ integrates Iin and produce the gate voltage of M6 shown in Fig. 5.5. V_τ sets the time constant, and V_g controls the gain of the integration [65, 67, 73–75].

The differential-pair integrator (DPI) synapse is shown in Fig. 5.5 to analyze the behavior of these synapses we can determine the following current equations [65, 67, 83].

$$I_{th} + I_C = I_{in} \tag{5.1.8}$$

$$I_C = I_{in} \times \frac{I_C}{I_{in}} = I_{in} \times \frac{I_C}{I_C + I_{th}} = \frac{I_{in}}{1 + \frac{I_{th}}{I_C}} \tag{5.1.9}$$

where Ith and IC are sub-threshold currents controlled by the gate voltage of M3, Vg, and the gate voltage of M4, V_C in Fig. 5.5.

The sub-threshold currents can be described in the following Eq. 5.1.10 [65, 67, 83].

$$I_C = I_0 e^{\frac{\kappa V_C}{U_T}}, \; I_{th} = I_0 e^{\frac{\kappa V_g}{U_T}} \tag{5.1.10}$$

$$\frac{I_{th}}{I_C} = e^{\frac{\kappa(V_g - V_C)}{U_T}} \tag{5.1.11}$$

$$I_C = \frac{I_{in}}{1 + e^{\frac{\kappa(V_g - V_C)}{U_T}}} \tag{5.1.12}$$

The output current, I_{out}, and I_g and I_C can be defined as a function of I_{out} and I_g as follows Eq. 5.1.13 [65, 67, 83].

$$I_{out} = I_0 e^{-\frac{\kappa(V_C - V_{DD})}{U_T}} \tag{5.1.13}$$

$$I_g = I_0 e^{-\frac{\kappa(V_g - V_{DD})}{U_T}} \tag{5.1.14}$$

$$I_C = \frac{I_{in}}{1 + \frac{I_{out}}{I_g}} \tag{5.1.15}$$

where I_g is the drain current of a PMOS transistor with the gate voltage of V_g, and I_{out} is shown in Fig. 5.5.

The derivative of the output current can be defined as follows Eq. 5.1.16 [65, 67, 83].

$$I_{\text{out}} = I_0 e^{-\frac{\kappa(V_C - V_{DD})}{U_T}} \tag{5.1.16}$$

$$\frac{d}{dt} I_{out} = I_0 e^{-\frac{\kappa(V_C - V_{DD})}{U_T}} \times \frac{d}{dt}\left(-\frac{\kappa(V_C - V_{DD})}{U_T}\right) \tag{5.1.17}$$

$$\frac{d}{dt} I_{\text{out}} = I_{\text{out}} \times \left(-\frac{\kappa}{U_T}\right)\frac{d}{dt}V_C \tag{5.1.18}$$

where $d/dt\,V_C$ can be determined from $Csyn$ current is shown in Fig. 5.5 [65, 67, 83].

$$C_{\text{syn}} \frac{d}{dt}V_C = I_\tau - I_C \Rightarrow \frac{d}{dt}V_C = \frac{I_C - I_\tau}{C_{\text{syn}}} \tag{5.1.19}$$

The resulting differential equation are as follows Eq. 5.1.20

$$\frac{d}{dt} I_{\text{out}} = \left(\frac{\kappa I_\tau}{U_T C_{\text{syn}}}\right)\left(\frac{\frac{I_{\text{in}}}{I_\tau}}{1 + \frac{I_{\text{out}}}{I_g}} - 1\right) \times I_{\text{out}} \xrightarrow{\text{if } \tau = \frac{U_T C_{\text{syn}}}{\kappa I_\tau}} \tau \frac{d}{dt} I_{\text{out}} + I_{\text{out}} = \frac{\frac{I_{\text{in}}}{I_\tau}}{\frac{1}{I_{\text{out}}} + \frac{1}{I_g}} \tag{5.1.20}$$

where τ is the time-constant and If $V_g > V_C$, the differential equation of the filter can be described as a first-order low-pass filter differential equation as follows Eq. 5.1.21

$$\tau \frac{d}{dt} I_{\text{out}} + I_{\text{out}} = \frac{I_{in} I_g}{I_\tau} \tag{5.1.21}$$

The output current is can be determined as follows (Fig. 5.5)

$$\text{rise} \Rightarrow I_{\text{out}}(t) = \frac{I_{\text{in}} I_g}{I_\tau}\left(1 - e^{-\frac{t - t^+}{\tau}}\right) + I_{\text{out}}\left(t^+\right)e^{-\frac{t - t^+}{\tau}} \tag{5.1.22}$$

$$\text{decay} \Rightarrow I_{\text{out}}(t) = I_{\text{out}}\left(t^+\right)e^{-\frac{t - t^+}{\tau}} \tag{5.1.23}$$

where $I_{out}(t^+)$ is the output current value in spike arrival time. The time constant of this filter depends on the value of the capacitor and V_τ. In this filter's circuit the values of V_g and V_w control the filter gain [65, 67, 83].

Combining a Winner-Take-All (WTA) circuit with a Differential Pair Integrator Synapse and a Leaky Integrate-and-Fire (LIF) neuron can create a sophisticated neural networks architecture as shown in Fig. 5.2. Differential pair integrator synapse with current source circuit is shown in Fig. 5.6. Winner-Take-All (WTA) circuit can be implemented with connecting to or more current conveyers as shown in Fig. 5.7. The WTA circuit select the synapse with the highest input signal strength wins and suppresses the activity of the other synapse [65, 67, 83].

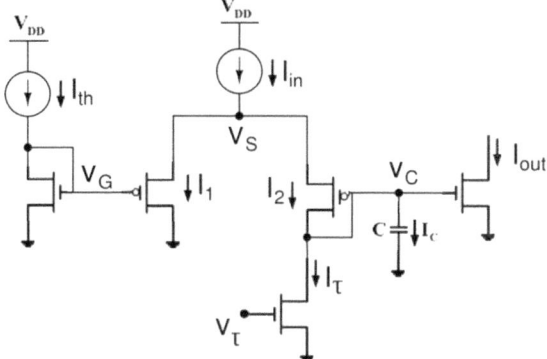

Fig. 5.6 Differential pair integrator synapse with current source circuit

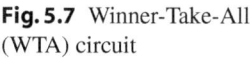

Fig. 5.7 Winner-Take-All
(WTA) circuit

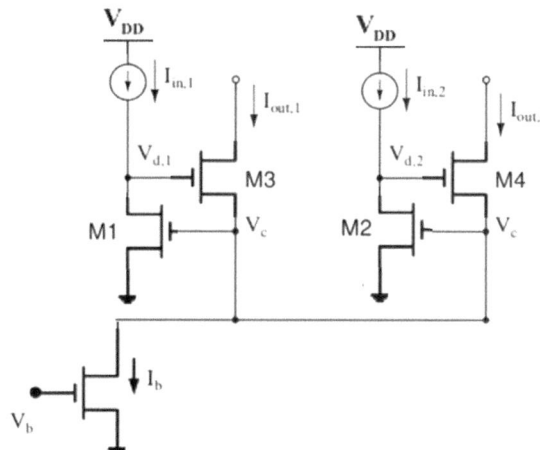

The input receives the analog vector input signals. The weighted multiplication multiply the input vector elements by corresponding weights in the analog vector matrix multiplication (VMM). The summation sum up the weighted products to get the final output [84–86]. The analog vector matrix multiplication input is fed to and is connected to the Winner Take All (WTA) circuit. The output provides the result of the vector matrix multiplication for further processing in the neural networks. The analog VMM circuits shown in Fig. 5.8 can process multiple elements of the input vector simultaneously enabling parallel computation [84–86].

The STDP is located in the learning plasticity module of the neural networks system as shown in Fig. 5.2 [65, 67, 76, 83, 87]. The STDP circuit adjust the synaptic weights based on spike timing. The capacitor is used for comparing spike timings. The STDP is typically located after the DPI synapse, WTA circuit, and LIF neuron in the block diagram as shown in Fig. 5.2.

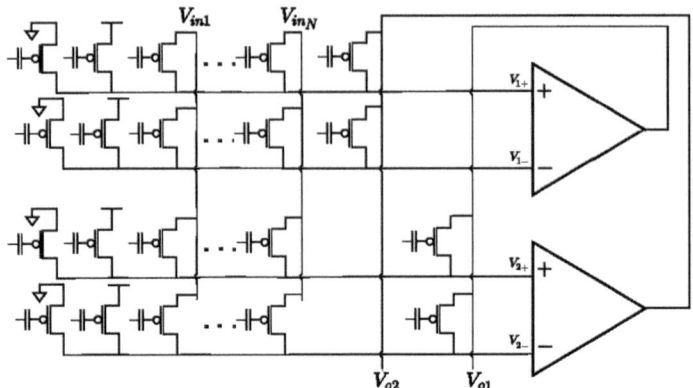

Fig. 5.8 Analog vector matrix multiplication circuit

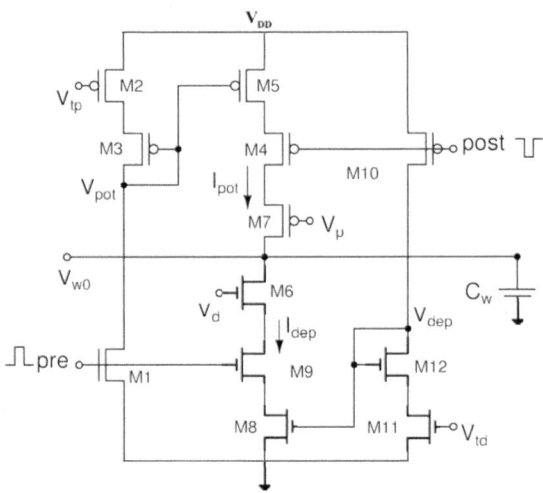

Fig. 5.9 STDP Spike-Timing Dependent Plasticity circuit

The Spike-Timing Dependent Plasticity (STDP) circuit shown in Fig. 5.9 detects spikes from the pre-synaptic neuron. It triggers when the pre-synaptic voltage crosses a threshold. The STDP measures the timing difference between the pre-synaptic and post-synaptic spikes. Then integrates the synapse current and adjusts the synaptic weight based on the timing information [65, 67, 76, 83, 87].

A Silicon Synapse with an Adaptive CMOS Neuron for Neuromorphic Computing

6

6.1   A CMOS Synapse Circuit Model for Spiking Neural Networks

Synapses are responsible for connecting neurons and communicating spike signals between them. A synapse receives spike voltages from the output of its pre-synaptic neuron. It produces a current based on a weight value. Then it feeds this weighted current to its post-synaptic neuron [32, 65–71, 73–75, 78, 79]. A CMOS synapse circuit model for Spiking Neural Network (SNN) is shown in Fig. 6.1. The CMOS synapse circuit design ensure that all transistors are working at the subthreshold level [30, 76, 77].

A CMOS synapse circuit model for Spiking Neural Networks (SNN) time-constant τ_d can be described by following equation [30, 76, 77]

$$\tau_d = \frac{C V_T}{k_n I_\tau} \tag{6.1.1}$$

where τ_d is the synapse time-constant and V_T is the thermal voltage, C is the capacitor $C1$ in synapse circuit shown in Fig. 6.1, kn is the transistor width to length ratio parameter and I_τ the sub threshold drain current of transistor $M7$ in Fig. 6.1.

Figure 6.1 shows the CMOS synapse circuit model for Spiking Neural Networks (SNNs). The design ensure that all transistors are working at the subthreshold level similar to the model provided by Horiuchi [65]. In the circuit shown in Fig. 6.1, the current through transistor M7 was the determinant factor to control the time-constant parameter. The CMOS synapse circuit model for Spiking simulation results is shown in Fig. 6.2.

Using multiple-gated transistors configuration M8 and M9 transistors, an additional branch parallel to transistor M7 provided additional control where the synaptic output time-constant can be tuned. Figure 6.3 shows the proposed modified adaptive CMOS synapse circuit SNN model with time-constant parameter tuning.

© The Author(s), under exclusive license to Springer Nature Switzerland AG 2025
Z. El-Khatib and S. Moussa, *Wireless Communication Using Deep Learning Techniques for Neuromorphic VLSI Computing*, Synthesis Lectures on Engineering, Science, and Technology, https://doi.org/10.1007/978-3-031-73800-5_6

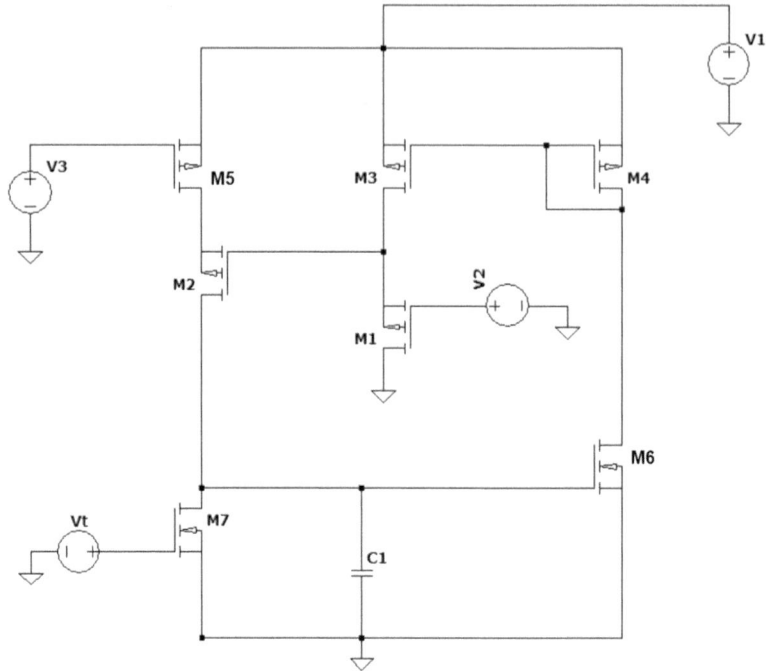

Fig. 6.1 A CMOS synapse circuit model for SNN

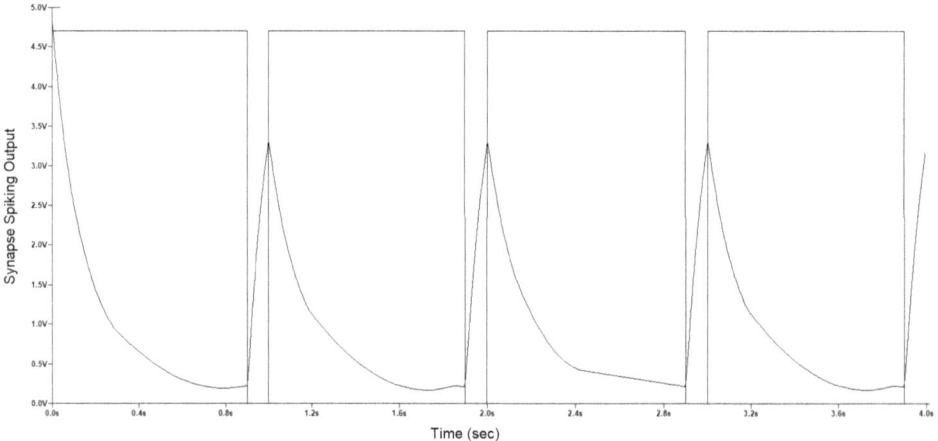

Fig. 6.2 The CMOS synapse circuit model for Spiking simulation results

Figure 6.1 shows the CMOS synapse circuit model for Spiking Neural Networks (SNNs). The design ensure that all transistors are working at the subthreshold level similar to the model provided by Horiuchi [65].

In the circuit shown in Fig. 6.1, the current through transistor M7 shown in Fig. 6.1 was the determinant factor to control the time-constant parameter. Using multiple gated transistors configuration M8 and M9 transistors, an additional branch parallel to transistor M7 provided additional control where the synaptic output time-constant can be tuned. Figure 6.3 shows the modified CMOS synapse circuit SNN model with time-constant parameter tuning.

The transistors are operating at the subthreshold level weak inversion region. At this mode of operation, the current-voltage relationship is exponential where $VT = kTq$ is the thermal voltage and $S = Wn/Ln$ is the transistor width to length ratio parameter and C1 is the capacitor in synapse circuit shown in Fig. 6.3.

Using Kirchoff's current law we get the following equation [30, 76, 77]

$$i_c = i_2 - i_\tau \qquad (6.1.2)$$

knowing that [30, 76, 77]

$$i_c = C \frac{dV_c(t)}{dt} \qquad (6.1.3)$$

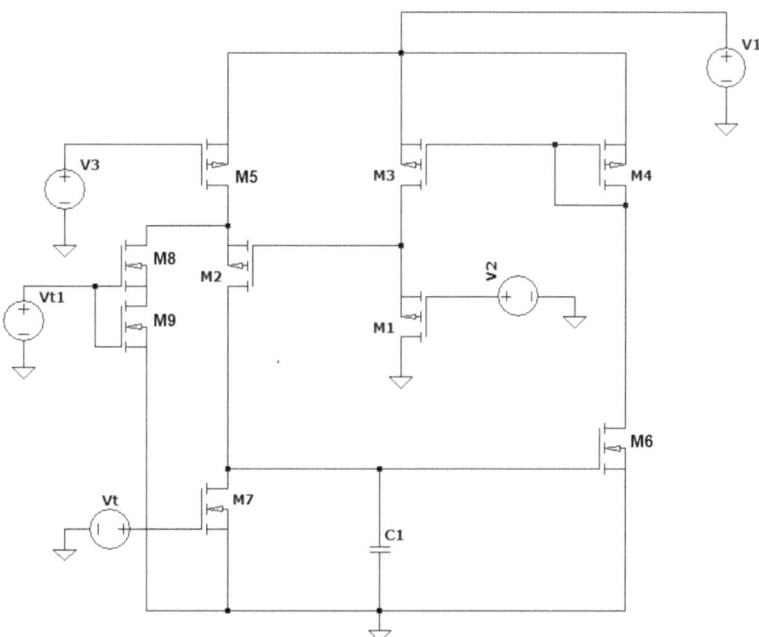

Fig. 6.3 The modified adaptive CMOS synapse circuit SNN model with time-constant tuning

Substituting Eqs. 6.1.3 in 6.1.2 we get [30, 76, 77]

$$C\frac{dV_c(t)}{dt} = SI_o e^{\frac{kp}{V_T}(V_{dd}-V_w)} - I_\tau \tag{6.1.4}$$

Knowing I_τ the sub-threshold drain current of transistor M7 in Fig. 6.3 can be determined by Eq. 6.1.5 [30, 76, 77]

$$I_\tau = S_7 I_o e^{\frac{V_T}{V_T}} k_n \tag{6.1.5}$$

$$i_{synapse}(t) = S_8 I_o e^{\frac{kn}{V_t} v_c(t)} \tag{6.1.6}$$

Taking the derivative of $I_{synapse}$ we get [30, 76, 77]

$$\frac{d i_{synapse}(t)}{dt} + \frac{k_n I_\tau}{CV_T} i_{synapse(t)} = \frac{S_8 k_n I_o}{CV_T} e^{\frac{kp}{V_T}(V_{dd}-V_w)} \tag{6.1.7}$$

The synapse time constant τ can be determined by [30, 76, 77]

$$\tau = \frac{CV_T}{K_n I_\tau} \tag{6.1.8}$$

where τ is the synapse time-constant and V_T is the thermal voltage, C is the capacitor C1 in synapse circuit shown in Fig. 6.3, kn is the transistor width to length ratio parameter and I_τ the sub threshold drain current of transistor $M7$ in Fig. 6.3.

The modified CMOS synapse circuit time-constant is given by [30, 76, 77]

$$\tau_r = \frac{CV_T}{k_n(I_{\tau r} + I_\tau)} \tag{6.1.9}$$

where τ_r is the modified synapse circuit time-constant and V_T is the thermal voltage, C1 is the capacitor in synapse circuit shown in Fig. 6.3, kn is the transistor width to length ratio parameter, I_τ the sub-threshold drain current of transistor $M7$ and $(I_\tau + I_\tau)$ is the total sub-threshold drain current of the multiple-gated transistor $M8$, $M9$ and $M7$ shown in Fig. 6.3. By using multiple-gated transistors configuration $M8$ and $M9$ transistors, an additional branch parallel to transistor $M7$ provide additional control where the synaptic output time-constant is tuned.

Table 6.1 shows the modified CMOS synapse circuit Transistor dimensions width and length.

The simulation results are shown in Fig. 6.4 illustrate the tuning of the modified circuit time-constant parameter. By using multiple-gated transistors configuration M8 and M9 transistors, an additional branch parallel to transistor M7 shown in Fig. 6.3 provide additional control where the synaptic output time-constant is tuned. The effect of changing multiple-gated transistors bias voltage on the decaying time-constant from $Vt1 = 0.25$ V bias voltage to $Vt1 = 0.3$ V and to $Vt1 = 0.45$ V is shown in Fig. 6.4. By using multiple-gated transistor configuration in the modified CMOS synapse the synaptic output current

Table 6.1 Adaptive modified CMOS synapse transistor dimensions

Transistor dimensions	Length	Width
M1	$2u$	$4u$
M2	$5u$	$10u$
M3	$5u$	$15u$
M4	$5u$	$10u$
M5	$5u$	$10u$
M6	$5u$	$10u$
M7	$5u$	$10u$
M8	$5u$	$15u$
M9	$5u$	$15u$

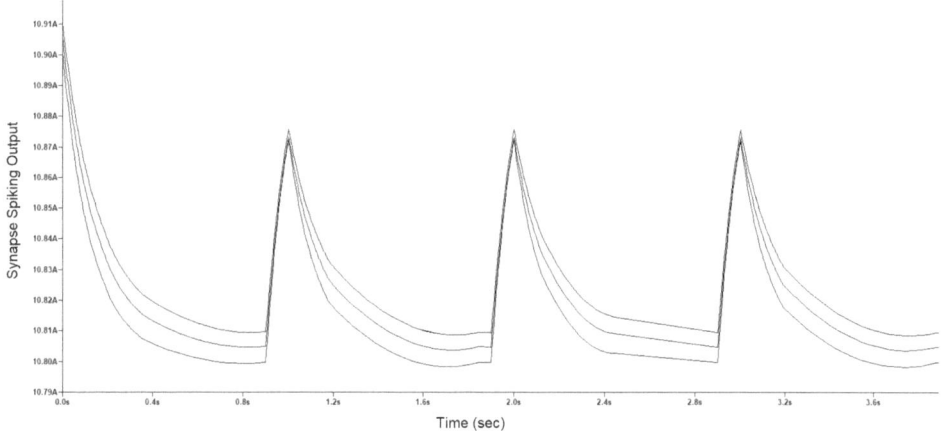

Fig. 6.4 The modified CMOS synapse circuit time-constant simulation results

time-constant is tuned. The effect of changing the multiple-gated transistor bias voltage from 0.25 to 0.45 V tunes the spiking output current exponential time-constant range by 200 ms as shown in simulation results in Fig. 6.4. By tuning the decaying exponential time-constant with multiple-gated transistor configuration, the proposed modified CMOS synapse captures the dynamic nature of biological synapses.

Merolla [26] synapse design with feedback control achieved a tunable time-constant range is 100 ms. Chen [27] design a memristor-based synapse with tunable time-constant range of 100 ms. Kim [28] synapse design with floating-gate has a 1 ms range of time-constant tuning. Tete [29] synapse design with varying capacitors achieved a tunable time-constant range of 10 us–100 ms. Liu [30] design a memristor-based synapse with tunable time-constant range of 1–100 ms. Hong [31] design a memristor-based synapse with tunable time-constant range of 100 us–100 ms. Table 6.2 shows the tunable of time-constant range

Table 6.2 Synapse design tunable time-constant range comparison

Synapse design	Circuit configuration	Time-constant tuning range
Merolla [26]	Feedback control	100 ms
Chen [27]	Memristor-based	100 ms
Kim [28]	Floating-gate	1 ms
Tete [29]	Varying capacitors	10 us–100 ms
Liu [30]	Memristor-based	1–100 ms
Hong [31]	Memristor-based	100 us–100 ms
This work	Multiple-gated	200 ms

of previously published synapse designs in comparison to the proposed synapse design. Our proposed synapse design with multiple-gated transistor configuration achieved a tunable time-constant range of 200 ms compared to previously published work with limited tunable time-constant range to 100 ms.

6.2   Spiking Integrate-and-Fire CMOS Neuron

A spiking integrate-and-fire CMOS neuron is a type of artificial neuron that is designed to simulate the behavior of biological neurons using complementary metal-oxide-semiconductor (CMOS) technology [65]. The integrate-and-fire CMOS neuron circuits can be used in various applications such as neural network systems, neuromorphic computing [63, 88–92] and brain-inspired computational systems [93–95]. A tunable spiking quadratic integrate-and-fire neuron incorporates a quadratic function to model the non-linear behavior of biological neurons more accurately than integrate-and-fire CMOS neuron. Neuromorphic circuits, including quadratic integrate-and-fire (QIF) CMOS neurons, have gained significant interest in the field of artificial intelligence and neuroscience [96, 97]. Due to their potential for high-speed, low-power, and parallel information processing, that makes them more efficient compared with Von Neumann bottleneck architecture [64, 80–82, 98–103].

The quadratic integrate-and-fire (QIF) CMOS neurons are typically implemented using CMOS technology. QIF CMOS neuron can be used in various applications, including spiking neural networks (SNNs), neuromorphic computing and parallel computing architectures such as brain-machine interfaces [91, 104–107]. By accurately modeling the behavior of biological neurons, the spiking integrate-and-fire neuron can enable precise control of assistive technologies [28, 108–110]. Indiveri and Horiuchi [65] implemented an integrate-and-fire CMOS neuron using differential pair topology achieving low power consumption. Indiveri and Horiuchi [65] implemented his integrate-and-fire neuron in 800 nm CMOS process using 20 transistors. Srinivasan and Cowan [132] implemented his CMOS neuron using current-mode circuit topology with limited tunability capability. Qiao [93] also designed his CMOS neuron using current-mode topology. Whereas Yu [108] implemented his neuron design using switch-capacitor circuit configuration. Wijekoon [111] designed a Quadratic Integrate-and-Fire CMOS neuron in 350 nm process using 14 transistors. Sourikopoulos [112] implemented an integrate-and-fire CMOS neuron in 65 nm process using 10 transistors. Whereas van Schaik [98] designed an Izhikevich CMOS neuron model in 90 nm process using 17 transistors. Liu [113] implemented his CMOS neuron using transconductance amplifier topology however it does not have circuit tunable capability. Moreover, Indiveri and Horiuchi [65] designed his integrate-and-fire neuron with membrane potential range of 1.5 V in 800 nm CMOS process however it does not have a tuning capability. Srinivasan and Cowan [132] designed his izhikevich neuron with membrane potential range of 150 mV in 65 nm CMOS process with a frequency tuning range of 200 Hz. Ou et al. [60] designed his Morris-Lecar neuron with membrane potential range of 200 mV in 180 nm CMOS process with a frequency tuning range of 290 Hz.

6.2.1 Leaky Integrate-and-Fire CMOS Neuron Model

A leaky integrate-and-fire neuron model includes a leak term to the membrane potential that reflects the diffusion of ions that occurs through the membrane when some equilibrium is not reached in the cell [114, 115]. The Leaky integrate-and-fire neuron model consist of a capacitor C in parallel with a resistor R driven by a current I(t). The driving current is split into two components and is described by the following equation [61, 114–117]

$$I(t) = I_R - I_C \tag{6.2.1}$$

where I(t) is the input current injected to the neuron and I_R is the current that flows through its membrane resistor and I_C is the current that flows through its circuit capacitor representing the neuron membrane capacitor.

Rearranging Eq. 6.2.1 with Ohms law and the capacitor current, we get the following equation.

$$I(t) = \frac{V_{\text{mem}}}{R} + C \cdot \frac{dV_{\text{mem}}}{dt} \tag{6.2.2}$$

Multiplying above Eq. 6.2.2 with R and introducing τ neuron membrane time constant, the above equation becomes [114, 115]

$$\tau \cdot \frac{dV_{\text{mem}}}{dt} = -V_{\text{mem}} + R \cdot I(t) \tag{6.2.3}$$

where V_{mem} is the neuron membrane potential.

Figure 6.5 shows a leaky integrate-and-fire CMOS neuron model including its circuit capacitor representing the neuron membrane capacitor.

The input current I_{in} injected to the neuron flowing through transistor M1 in Fig. 6.5 can be described as [61, 114–117]

Fig. 6.5 Modeling a Leaky Integrate-and-Fire CMOS neuron

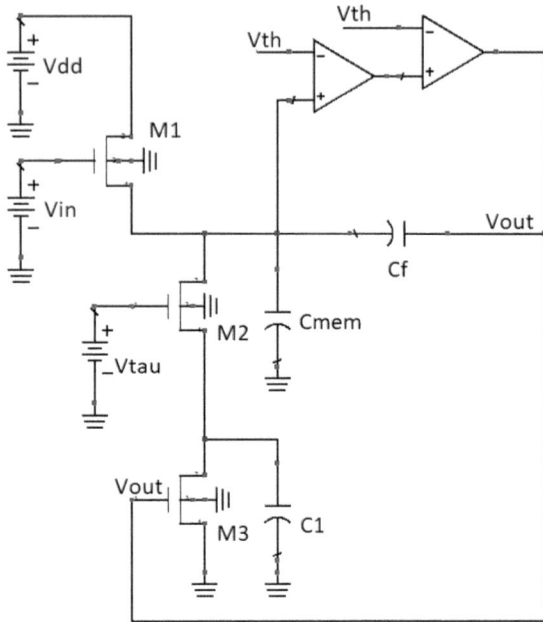

$$I_m = I_{th} e^{\frac{K(V_{dd} - V_T - V_{T0}))}{V_T}} \left(1 - e^{\frac{-(V_{dd} - V_{mem})}{V_T}} \right) \tag{6.2.4}$$

The leak current I_{leak} injected to the neuron flowing through transistor M2 in Fig. 6.5 can be described as [61, 114, 115]

$$I_{leak} = I_{th} \, e^{K(V_t - V_1 - V_{T0})} \left(1 - e^{\frac{-(V_{mem} - V_1)}{V_T}} \right) \tag{6.2.5}$$

Adding both currents by taking the current node equation between M1 and M2, the current flowing through the membrane capacitance can be described as

$$\left(\frac{dV}{dt} V_{mem} \right) (C_{mem} + C_f) = I_W - I_{leak} \tag{6.2.6}$$

In a leaky integrate-and-fire CMOS neuron model, the neuron will fire when the input current I exceeds the threshold current, otherwise it will leak out any potential change. It exhibits periodic spiking T [61, 116, 117],

$$T = \tau \ln \left(\frac{R I_o}{R I_o - V_{th}} \right) \tag{6.2.7}$$

Fig. 6.6 Matlab simulation of a leaky integrate-and-fire membrane potential with constant input current

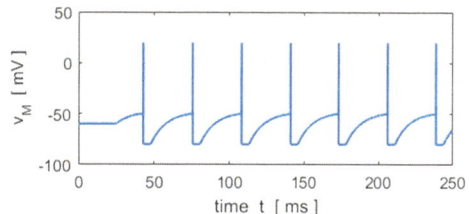

Fig. 6.7 Tensorflow python simulation of a leaky integrate-and-fire membrane potential with constant input current

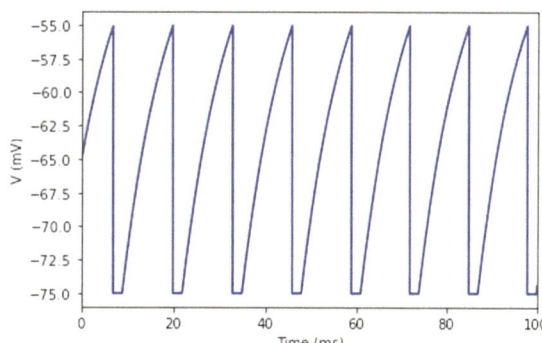

where τ is the time constant of the circuit and V_{th} is the threshold voltage. Since there is a spike every time the capacitance discharges, the spike firing frequency f is the reciprocal of time T and the spike firing frequency f can be described as

$$f = \frac{1}{\tau \ln\left(\frac{RI_O}{RI_O - V_{th}}\right)}. \tag{6.2.8}$$

Figure 6.6 shows a Matlab simulation of a leaky integrate-and-fire membrane potential with constant input current.

Figure 6.7 shows a Tensorflow python simulation of a leaky integrate-and-fire membrane potential with constant input current.

Figure 6.8 shows another implementation of a leaky integrate-and-fire subthreshold CMOS neuron circuit [65].

In Fig. 6.8, the current flowing through membrane capacitance C_v can be described as

$$\left(Cv\frac{dV}{dt}\right) = I_{M3} - I_{M5} + I_{syn} \tag{6.2.9}$$

Whereas the current flowing the circuit capacitance C_U can be described as

$$\left(C_U\frac{dV}{dt}\right) = I_{M4} - I_{M6}. \tag{6.2.10}$$

Fig. 6.8 A leaky
integrate-and-fire subthreshold
CMOS neuron implementation

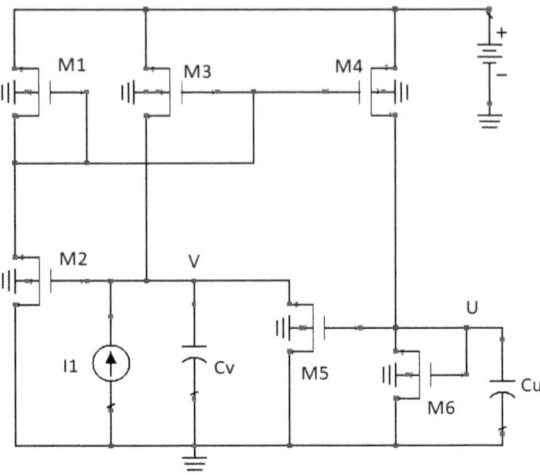

6.2.2   Quadratic Integrate-and-Fire CMOS Neuron Model

The quadratic integrate-and-fire neuron model is described by the following differential
equation [61, 63, 95, 116–118]

$$\frac{dV}{dt} = (V - V_{\text{reset}})(V - V_{\text{th}}) + I \tag{6.2.11}$$

where dV/dt is the rate of change of the membrane potential and V is the membrane
potential. $Vreset$ is the reset potential. Vth is the threshold potential, and I is the input
current [61, 63, 95, 116–118].

The quadratic integrate-and-fire neuron model can be rearranged into the following equation. R and C being the resistance and capacitance respectively of the quadratic integrate-
and-fire neuron circuit integrator [61, 115, 118]

$$RC\frac{d}{dt}V = V^2 + I \tag{6.2.12}$$

The quadratic term captures the neuron firing rate nonlinear quadratic behavior. The
quadratic integrate-and-fire neuron model can be implemented as an analog circuit implementation where V is the voltage across the cell membrane and I is the input current with a
given reset condition described as follows, [70, 115, 117–119]

$$V = \frac{1}{RC} \int V^2 + I \text{ When } V > V_{\text{peak}} \text{ then } V = V_{\text{reset}} \tag{6.2.13}$$

The quadratic integrate-and-fire neuron model exhibits a periodic spiking T that can be
described as follows [61, 116–118, 120]

$$T = \frac{1}{2\sqrt{I}} \ln\left(\frac{\left(V_{\text{peak}} - \sqrt{I}\right)\left(V_{\text{reset}} + \sqrt{I}\right)}{\left(V_{\text{peak}} + \sqrt{I}\right)\left(V_{\text{reset}} - \sqrt{I}\right)}\right) \tag{6.2.14}$$

where the Vpeak is the peak membrane potential and Vreset is the reset potential, and I is the input current.

6.3   Proposed Adaptive Quadratic Integrate-and-Fire CMOS Neuron

The proposed adaptive quadratic integrate-and-fire (AQIF) CMOS neuron is shown in Fig. 6.9. A differential pair with variable capacitor integrator, a tunable schmitt trigger threshold detector circuit and a switch MOSFET transistor are integrated in the adaptive quadratic integrate-and-fire CMOS neuron which models the quadratic neuron behavior. The proposed AQIF CMOS neuron has the ability to adjust the spiking frequency without changing the input current.

The quadratic integrate-and-fire CMOS neuron dynamics is represented in the form of a differential equation describing the action potential of the neuron as [68, 69, 83, 117, 118, 120, 121]

$$C_m \frac{dV}{dt} = g_L \frac{(V - V_{\text{th}})(V - V_{\text{reset}})}{(V_{\text{th}} - V_{\text{reset}})} + I \tag{6.3.1}$$

where C_m is the membrane capacitance, V_{th} is the threshold potential, V_{reset} is the reset potential, I is the input current and g_L is the leak conductance. The rate of change of the membrane potential dV/dt is determined by the input current I, leak conductance g_L, and the equation quadratic term $((V - V_{th})(V - V_{reset}))/((V_{th} - V_{reset}))$. This quadratic term introduces nonlinearity into the neuron's dynamics resulting in its spiking behavior.

The adaptive AQIF CMOS neuron membrane time constant τ, near threshold voltage, can be described as [62, 65, 68, 69, 118, 121, 122]

$$\tau = \frac{C(V_{\text{th}} - V_{\text{reset}})}{g_L(V_{\text{th}} + V_{\text{reset}})} \tag{6.3.2}$$

The adaptive AQIF CMOS neuron exhibits firing rate f as a function of input current I to the neuron [71, 89, 114, 117, 123–127]

$$f = \frac{2\sqrt{I}}{\tau \cdot (V_{\text{th}} - V_{\text{reset}}) \cdot \ln\left(\frac{(V_{\text{th}} - V_{\text{mem}}) + \sqrt{I}}{(V_{\text{reset}} - V_{\text{mem}}) + \sqrt{I})}\right)} \tag{6.3.3}$$

where τ is the neuron membrane time constant and R is the membrane resistance and C is the neuron membrane capacitance. V_{mem} represents the membrane potential of the AQIF CMOS neuron, V_{th} is the threshold potential at which the neuron fires an action potential,

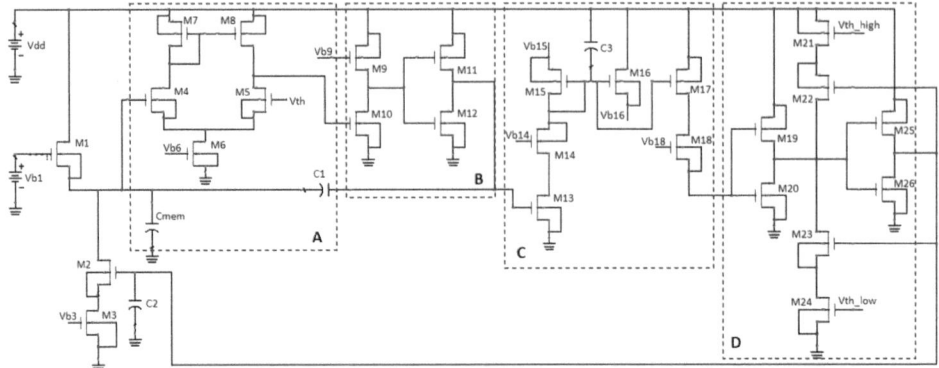

Fig. 6.9 Proposed adaptive quadratic integrate-and-fire CMOS neuron **A** differential pair integrator **B** voltage amplifier **C** variable diode capacitor **D** tunable schmitt trigger

V_{reset} is the reset potential after a spike occurs, I is the input current to the neuron, and τ is the time constant associated with the neuron membrane potential [66, 67, 128–131].

The proposed adaptive quadratic integrate-and-fire CMOS neuron shares the ability to adjust its parameters to control its behavior and response characteristics. The differential amplifier integrator (M4–M8) circuit is coupled with a variable diode capacitor (M13–M18) circuit as shown in Fig. 6.9. The variable diode capacitor is placed between the integration stage and the tunable Schmitt trigger (M19–M26) circuit. The positive feedback occurs through the quadratic nonlinearity introduced by the nonlinear element switch MOSFET M2 operating in the saturation region to implement the quadratic dynamics of the CMOS neuron.

The integrated output voltage, which represents the membrane potential, is fed into the variable diode capacitor. The variable diode capacitor is a circuit that uses a diode transistor M15 and a capacitor C3 of 1nF in conjunction. The diode transistor M15 acts as a variable resistor, allowing the time constant to be changed by varying the diode transistor bias voltage.

The membrane time constant $\tau = (C(V_{th} - V_{reset}))/(g_L(V_{th} + V_{reset}))$ in Eq. 6.2.14 the parameter that can be varied by the variable diode capacitor is the leak conductance g_L. The variable diode capacitor circuit allows for the adjustment of the time constant τ by changing the bias voltage of the diode transistor M15. In the Eq. 6.2.14, the membrane time constant τ is inversely proportional to the leak conductance g_L.

By adjusting the bias voltage of the diode in the variable diode capacitor integrator circuit, the effective leak conductance can be adjusted, resulting in a corresponding change in the time constant of the circuit. Therefore, by varying the diode bias voltage V_{bias} of the variable diode capacitor integrator, the time constant τ can be tuned to achieve the desired spike frequency behavior of the circuit as shown in Eq. 6.3.4.

$$V_{bias} = \frac{\tau g_L (V_{th} + V_{reset})}{C + V_{th} - V_{reset}} \qquad (6.3.4)$$

Table 6.3 Variable capacitor integrator bias voltage tuning for spike frequency 312.5 Hz

Parameter	
Membrane time constant τ	3.2 ms
Leak conductance g_L	100 μS
Threshold potential V_{th}	0.35 V
Reset potential V_{reset}	-0.75 V
Bias voltage V_{bias}	1.1 V

Table 6.4 Variable capacitor integrator bias voltage tuning for spike frequency 58.4 Hz

Parameter	
Membrane time constant τ	17.1 ms
Leak conductance g_L	100 μS
Threshold potential V_{th}	0.35 V
Reset potential V_{reset}	-0.75 V
Bias voltage V_{bias}	0.9 V

Therefore, to achieve a time constant of 3.2 ms corresponding to spike frequency 312.5 Hz, the variable diode capacitor bias voltage in the circuit is adjusted to 1.1 V as shown in Table 6.3.

To achieve a time constant of 17.1 ms corresponding to spike frequency 58.4, the variable diode capacitor bias voltage in the circuit is adjusted to 0.9 V as shown in Table 6.4.

The output of the variable diode capacitor is passed to the tunable Schmitt trigger (M19–M26) circuit, which compares the voltage with adjustable threshold levels to generate the output spike when the upper threshold is reached. The tunable Schmitt trigger circuit is set with the upper threshold $V(th - high)$ to 0.35 V and the lower threshold $V(th - low)$ is set to -0.2 V as shown in Table 6.5. The membrane potential gradually increases due to

Table 6.5 Tunable CMOS Schmitt trigger parameters

Parameter	
Upper threshold potential V_{th} high	0.35 V
Lower threshold potential V_{th}	-0.2 V
Reset potential V_{reset}	-0.75 V

the injected current. It continues to rise until it reaches the upper threshold $V(th - high)$. When the membrane potential V exceeds the upper threshold $V(th - high)$ of 0.35 V, the Schmitt trigger CMOS circuit detects this and generates a positive spike or action potential. After generating the spike, the membrane potential is reset to the reset potential V_{reset} of −0.75 V to simulate the refractory period of the neuron.

Adaptive Quadratic Integrate-and-Fire CMOS Neuron Performance

7

7.1 Adaptive Quadratic Integrate-and-Fire CMOS Neuron Simulation Performance

The membrane potential starts to integrate incoming currents again until it reaches the lower threshold $V_{(th-low)}$ of -0.2 V. As the membrane potential gradually increases, the Schmitt trigger circuit detects the crossing of $V_{(th-low)}$ of -0.2 V and generates another positive spike indicating another action potential. As the membrane potential crosses the upper threshold, this process repeats generating action potentials at regular intervals as shown in Figs. 7.1 and 7.2.

Figure 7.1 shows the simulation results for the proposed adaptive QIF neuron circuit quadratic voltage behavior. Figure 7.2 shows simulation results for the proposed adaptive QIF neuron circuit spiking voltage and with 58.4 Hz spiking with The Vpeak = 0.95 V and Vreset = -0.75 V and a switching threshold of 0.35 V.

The tunability of the proposed adaptive QIF CMOS neuron circuit refers to the ability to adjust its parameters to control its behavior characteristics. Figure 7.2 shows simulation results for the proposed adaptive QIF CMOS neuron circuit spiking voltage with 58.4 Hz spiking frequency and 17.1 ms spiking period.

Figure 7.3 shows simulation results for the proposed adaptive QIF neuron circuit Spiking behavior FFT with 58.4 Hz spiking frequency.

Z. El-Khatib and S. Moussa, *Wireless Communication Using Deep Learning Techniques for Neuromorphic VLSI Computing*, Synthesis Lectures on Engineering, Science, and Technology, https://doi.org/10.1007/978-3-031-73800-5_7

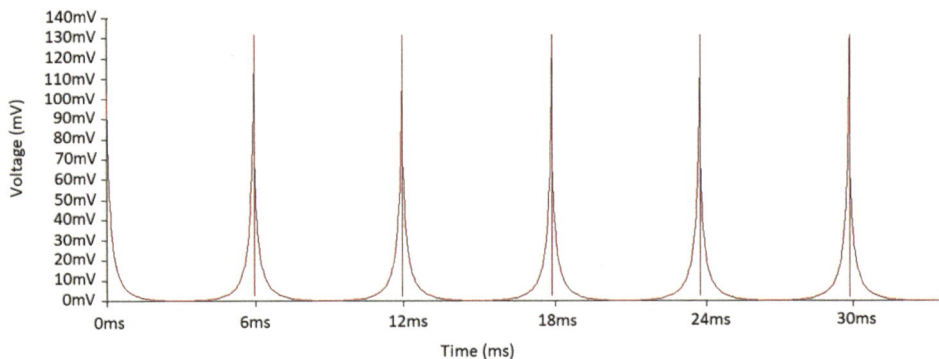

Fig. 7.1 Simulation results for the proposed adaptive AQIF neuron circuit quadratic voltage behavior

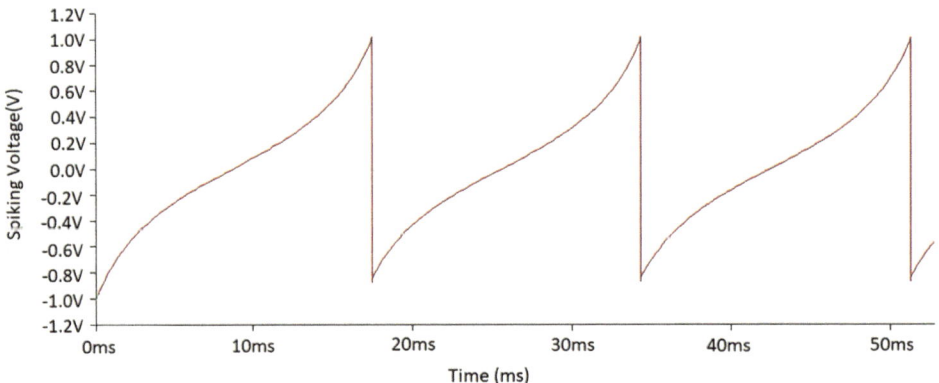

Fig. 7.2 Simulation results for the proposed adaptive AQIF neuron circuit spiking voltage with 58.4 Hz spiking frequency and 17.1 ms spiking period

Figure 7.4 shows simulation results for the proposed adaptive QIF neuron circuit spiking voltage with 312.5 Hz spiking frequency and 3.2 ms spiking period.

Figure 7.5 shows simulation results for the proposed adaptive QIF neuron circuit spiking behavior FFT with 312.5 Hz spiking frequency.

The fully integrated AQIF circuit number of transistors is 26 transistors as shown in Table 7.1. Fully-integrated CMOS neuron comparison with the proposed AQIF neuron is shown in Table 7.1. The tunable spiking frequency CMOS neuron comparison with the proposed AQIF neuron is shown in Table 7.2. Tunability is defined as the ability to adjust spiking frequency without changing the input current I. As shown in Table 7.2, Indiveri and Horiuchi [65] designed his integrate-and-fire neuron with total of 20 transistors and with membrane potential range of 1.5 V in 800 nm CMOS process however it does not have a tuning capability. Indiveri and Horiuchi [65] implemented his Integrate-and-Fire CMOS

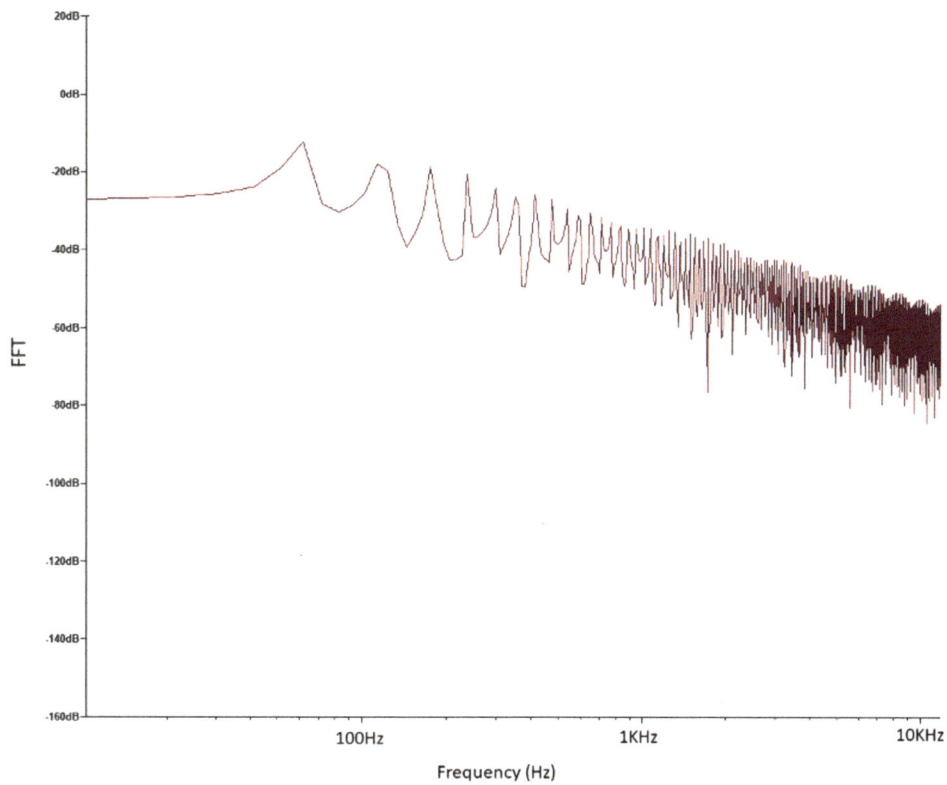

Fig. 7.3 Simulation results for the proposed adaptive QIF neuron circuit spiking behavior FFT with 58.4 Hz spiking frequency

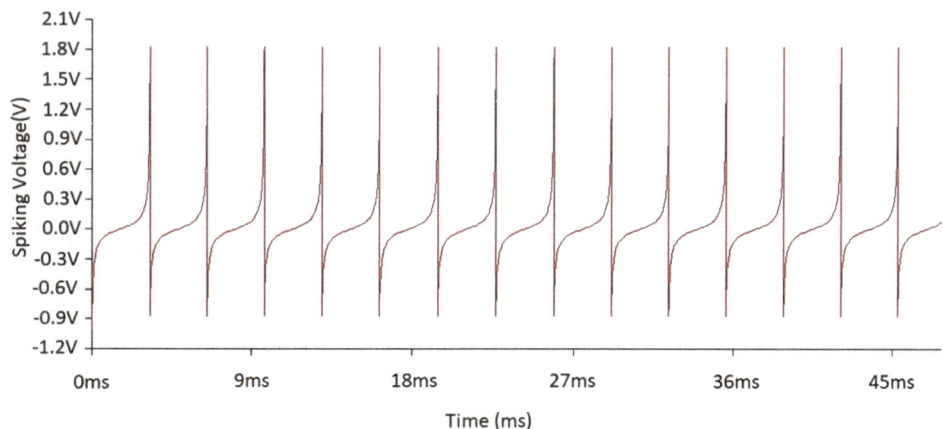

Fig. 7.4 Simulation results for the proposed adaptive QIF neuron circuit spiking behavior FFT with 312.5 Hz spiking frequency

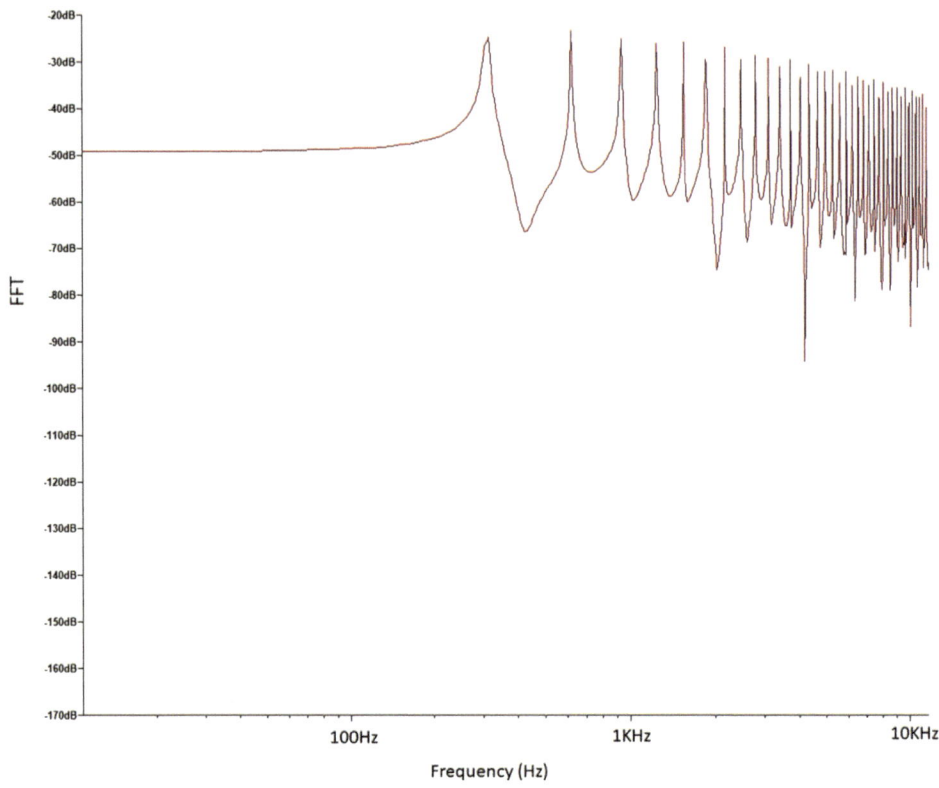

Fig. 7.5 Simulation results for the proposed adaptive QIF neuron circuit spiking behavior FFT with 312.5 Hz spiking frequency

Table 7.1 Fully-integrated CMOS neuron comparison with proposed AQIF neuron

	Neuron model	Transistor count	Energy/Spike [p]	CMOS process (nm)
Indiveri and Horiuchi [65]	Integrate-and-Fire	20	900	800
Wijekoon [111]	Quadratic I&F	14	790	350
Srinivasan and Cowan [132]	Izhikevich	18	1060	65
Sourikopoulos [112]	Integrate-and-Fire	10	720	65
Van Schaik [98]	Izhikevich	17	700	90
This work	AQIF	26	860	130

Table 7.2 Tunable Spiking frequency CMOS neuron comparison with proposed AQIF neuron

	Configuration	Frequency tuning range (Hz)	Membrane potential range	Tunability	Process (nm)
Indiveri and Horiuchi [65]	Integrate-and-Fire	0	1.5 V	None	800
Srinivasan and Cowan [132]	Izhikevich	200	150 mV	Tunable	65
Ou et al. [60]	Morris-Lecar	290	200 mV	Tunable	180
This work	AQIF	255	1.7 V	Tunable	130

neuron using differential pair topology and achieved low power consumption. Srinivasan and Cowan [132] designed his izhikevich neuron with limited membrane potential range of 150 mV in 65 nm CMOS process with a frequency tuning range of 200 Hz. Srinivasan and Cowan [132] did his CMOS neuron using current-mode circuit topology with total of 18 transistors as shown in Table 7.1 and with limited tunability capability. Van Schaik [98] designed his izhikevich neuron with total of 17 transistors. Indiveri and Horiuchi [65] did his quadratic integrate-and-fire neuron with 14 transistors however with no tuning capability. Whereas Ou et al. [60] designed a Morris-Lecar neuron with limited membrane potential range of 200 mV in 180 nm CMOS process with a frequency tuning range of 290 Hz as shown in Table 7.2.

Conclusion

8

8.1 Wireless Communications Using Deep Learning Models on Spiking Neural Networks Hardware

Automatic signal modulation recognition in AI-based wireless communication systems can be implemented using combinatorial deep learning neural network techniques to improve resource shortage and spectrum utilization efficiency for dynamic spectrum allocation. An automatic signal modulation classification model using combinatorial deep learning technique is presented. The proposed deep learning model increase accuracy for low signal to noise ratio SNR and maintain a high classification accuracy for high SNR signals. Using a hybrid deep learning model combining both ConvLSTM2D with Transformer-block neural networks, the proposed modulation classifier architecture can learn the signal for both low and high SNR and get better accuracy for signals with high noise.

Automatic signal modulation recognition in AI-based wireless communication can be done using combinatorial deep learning neural network techniques to improve resource shortage and spectrum utilization efficiency for dynamic spectrum allocation. An automatic signal modulation classification model using combinatorial deep learning technique was presented. Our proposed deep learning model increase accuracy for low signal to noise ratio SNR and maintain a high classification accuracy for high SNR signals. Using a hybrid deep learning model combining both ConvLSTM2D with Transformer-block neural networks, the proposed modulation classifier architecture can learn the signal for both low and high SNR and get better accuracy for signals with high noise. The proposed deep learning modulation classification technique achieves improved classification accuracy of 66% for low SNR signals and 93.5% at high SNR showing that our model is robust under noisy signal modulation. Our deep learning radio modulation classification model works using raw signal without the need of denoising the noisy signal. The simulation results show that the

© The Author(s), under exclusive license to Springer Nature Switzerland AG 2025
Z. El-Khatib and S. Moussa, *Wireless Communication Using Deep Learning Techniques for Neuromorphic VLSI Computing*, Synthesis Lectures on Engineering, Science, and Technology, https://doi.org/10.1007/978-3-031-73800-5_8

proposed technique outperforms existing feature-based extraction architectures in terms of modulation recognition performance getting better accuracy in lower SNR signals without sacrifice accuracy in higher SNR signals.

Spiking Neural Networks (SNN) is used to connect machine learning and neuroscience. Unlike Artificial Neural Networks (ANN), Spiking Neural Networks (SNN) do not fire continuously. The brain's energy efficiency for decision making cognitive tasks made scientists to focus their efforts on building non-Von Neumann computer systems that imitate the biological brain. Neurons process information as asynchronous event-driven spikes and retain memories as synaptic strengths of their connection in the brain. Analog VLSI is utilized to design spiking neural networks circuits such as silicon synapse and CMOS neuron. Because transistors have properties similar to nerve membrane channels. When transistors are operated in weak inversion region, they leak a very small current. This transistor region of operation is also known as the subthreshold region. This way a large network of thousands of neurons will consume very low power. Spiking Neural Network (SNN) do not fire continuously. SNN fires only when the post-synaptic potential reaches a certain threshold value making it suitable for low power design. Spiking neural network (SNN) provides a promising solution for low-power hardware for neuromorphic computing. Spiking neural network is more promising than other neural networks that can pave a new way for low-power neuromorphic computing applications.

Using deep learning neural network circuit methods and doing parallel computations on hardware can reduce costs. Spiking neural network (SNN) provides a promising solution for low-power hardware for neuromorphic computing. Spiking neural network is more promising than other neural networks that can pave a new way for low-power computing applications. The description and analysis of a silicon synapse circuit in Chap. 6. Synapses are responsible for connecting neurons and communicating spike signals between them. A synapse receives spike voltages from the output of its pre-synaptic neuron. It produces a current based on a weight value. Then it feeds this weighted current to its post-synaptic neuron. In Chap. 6 the design of a fully integrated adaptive modified CMOS synapse circuit is presented. Chapter 6 also presented an adaptive CMOS neuron For neuromorphic computing. The design a fully integrated adaptive quadratic integrate-and-fire CMOS neuron was presented. In Chap. 7 the adaptive quadratic integrate-and-fire CMOS neuron performance is presented.

Spiking neural network (SNN) provides a promising solution for low-power hardware for neuromorphic computing. Spiking neural network is more promising than other neural networks that can pave a new way for low-power neuromorphic computing applications. Spiking Neural Networks (SNN) is used to connect machine learning and neuroscience. Analog VLSI is utilized to design spiking neural networks circuits such as silicon synapse and CMOS neuron. The design of a fully integrated adaptive modified CMOS synapse circuit is presented. By using multiple-gated transistor configuration in the modified CMOS synapse an additional branch provide control where the synaptic output current time-constant is tuned. The effect of changing the multiple-gated transistor bias voltage from 0.25 to 0.45 V tunes the spiking output current exponential time-constant range as shown in simulation results. Our proposed synapse design with multiple-gated transistor configuration achieved a tunable time-constant range of 200 ms compared to previously published work with limited tunable time-constant range to 100 ms. By tuning the decaying exponential time-constant with multiple-gated transistor configuration, the proposed modified CMOS synapse captures the dynamic nature of biological synapses.

Moreover, the design of a fully integrated adaptive quadratic integrate-and-fire CMOS neuron was presented as well. A differential pair with variable capacitor integrator and a tunable schmitt trigger threshold detector circuit are integrated in the CMOS neuron that can be tuned varying its spiking frequency. The proposed adaptive QIF neuron has the ability to adjust the spiking frequency without changing the input current. The simulation results show the proposed CMOS neuron circuit spiking frequency can be tuned from 58.4 to 312.5 Hz and its spiking period from 17.1 to 3.2 ms with tuning the bias voltage of variable capacitor integrator. Having a peak voltage Vpeak = 0.95 V, a reset voltage Vreset = −0.75 V and a voltage threshold of 0.35 V with a membrane potential range of 1.5 V. The proposed CMOS neuron number of transistors is 26 designed in 130 nm process with a supply voltage of 1.8 V and a total power dissipation of 1.8 mW.

References

1. W. M. Lees, A. Wunderlich, P. Jeavons, P. D. Hale, and M. R. Souryal, "Deep learning classification of 3.5 ghz band spectrograms with applications to spectrum sensing," 6 2018.
2. A. Vaswani, N. Shazeer, N. Parmar, J. Uszkoreit, L. Jones, A. N. Gomez, L. Kaiser, and I. Polosukhin, "Attention is all you need," 6 2017.
3. V. Sathyanarayanan, J. Burke, R. Shang, and R. Bell, "Modulation classification using neural networks," 2019.
4. S. Huang, L. Chai, Z. Li, D. Zhang, Y. Yao, Y. Zhang, and Z. Feng, "Automatic modulation classification using compressive convolutional neural network," *IEEE Access*, vol. 7, pp. 79636–79643, 2019.
5. Y. Guo and X. Wang, "Modulation signal classification algorithm based on denoising residual convolutional neural network," *IEEE Access*, vol. 10, pp. 121733–121740, 2022.
6. R. Zhang, Z. Yin, Z. Wu, and S. Zhou, "A novel automatic modulation classification method using attention mechanism and hybrid parallel neural network," *Applied Sciences (Switzerland)*, vol. 11, pp. 1–19, 2 2021.
7. N. E. West and T. J. O'Shea, "Deep architectures for modulation recognition," 3 2017.
8. S. Chen, Y. Zhang, Z. He, J. Nie, and W. Zhang, "A novel attention cooperative framework for automatic modulation recognition," *IEEE Access*, vol. 8, pp. 15673–15686, 2020.
9. J. Jiang, Z. Wang, H. Zhao, S. Qiu, and J. Li, "Modulation recognition method of satellite communication based on cldnn model," pp. 1–6, IEEE, 6 2021.
10. B. Tang, Y. Tu, Z. Zhang, and Y. Lin, "Digital signal modulation classification with data augmentation using generative adversarial nets in cognitive radio networks," *IEEE Access*, vol. 6, pp. 15713–15722, 3 2018.
11. J. Xu, C. Luo, G. Parr, and Y. Luo, "A spatiotemporal multi-channel learning framework for automatic modulation recognition," *IEEE Wireless Communications Letters*, vol. 9, pp. 1629–1632, 10 2020.
12. K. Jiang, X. Qin, J. Zhang, and A. Wang, "Modulation recognition of communication signal based on convolutional neural network," *Symmetry*, vol. 13, 12 2021.

© The Editor(s) (if applicable) and The Author(s), under exclusive license to Springer
Nature Switzerland AG 2025
Z. El-Khatib and S. Moussa, *Wireless Communication Using Deep Learning Techniques for Neuromorphic VLSI Computing*, Synthesis Lectures on Engineering,
Science, and Technology, https://doi.org/10.1007/978-3-031-73800-5

13. Z. Liang, M. Tao, L. Wang, J. Su, and X. Yang, "Automatic modulation recognition based on adaptive attention mechanism and resnext wsl model," *IEEE Communications Letters*, vol. 25, pp. 2953–2957, 9 2021.

14. S. Chang, S. Huang, R. Zhang, Z. Feng, and L. Liu, "Multitask-learning-based deep neural network for automatic modulation classification," *IEEE Internet of Things Journal*, vol. 9, pp. 2192–2206, 2 2022.

15. W. Zhang, X. Yang, C. Leng, J. Wang, and S. Mao, "Modulation recognition of underwater acoustic signals using deep hybrid neural networks," *IEEE Transactions on Wireless Communications*, vol. 21, pp. 5977–5988, 8 2022.

16. B. Zou, X. Zeng, and F. Wang, "Research on modulation signal recognition based on cldnn network," *Electronics (Switzerland)*, vol. 11, 5 2022.

17. J. Bai, J. Yao, J. Qi, and L. Wang, "Electromagnetic modulation signal classification using dual-modal feature fusion cnn," *Entropy*, vol. 24, 5 2022.

18. X. Hao, Y. Luo, Q. Ye, Q. He, G. Yang, and C. C. Chen, "Automatic modulation recognition method based on hybrid model of convolutional neural networks and gated recurrent units," *Sensors and Materials*, vol. 33, 2021.

19. U. Dampage, S. M. Amarasooriya, R. A. Samarasinghe, and N. A. Karunasingha, "Combined classifier-demodulator scheme based on lstm architecture," *Wireless Communications and Mobile Computing*, vol. 2022, 2022.

20. K. Liu, W. Gao, and Q. Huang, "Automatic modulation recognition based on a dcn-bilstm network," *Sensors*, vol. 21, pp. 1–17, 3 2021.

21. H. Yang, L. Zhao, G. Yue, B. Ma, and W. Li, "Irlnet: A short-time and robust architecture for automatic modulation recognition," *IEEE Access*, vol. 9, pp. 143661–143676, 2021.

22. V. Kumaran, "Ensemble of deep learning enabled modulation signal classification model for underwater acoustic communication," 2022.

23. Z. Ke and H. Vikalo, "Real-time radio technology and modulation classification via an lstm auto-encoder," 11 2020.

24. C. Zhang, S. Yu, G. Li, and Y. Xu, "The recognition method of mqam signals based on bp neural network and bird swarm algorithm," *IEEE Access*, vol. 9, pp. 36078–36086, 2021.

25. T. K. Oikonomou, S. A. Tegos, D. Tyrovolas, P. D. Diamantoulakis, and G. K. Karagiannidis, "On the error analysis of hexagonal-qam constellations," *IEEE Communications Letters*, vol. 26, pp. 1764–1768, 8 2022.

26. P. A. Merolla and J. V. Arthur, "A million spiking-neuron integrated circuit with a scalable communication network and interface," *Science*, 2014.

27. M. Hu and Y. Chen, "A compact memristor-based dynamic synapse for spiking neural networks," *IEEE Transactions on Computer-Aided Design of Integrated Circuits and Systems*, 2017.

28. M. W. Kwon, H. Kim, J. Park, and B. G. Park, "Integrate-and-fire neuron circuit and synaptic device with floating body mosfets," *Journal of Semiconductor Technology and Science*, vol. 14, pp. 755–759, 12 2014.

29. M. Hu and Y. Chen, "Design of dynamic synapse circuits with vlsi design approach," *International Conference on Emerging Trends in Engineering and Technology*, 2010.

30. S. Liu, "Analog vlsi circuits for short-term dynamic synapses," *EURASIP Journal on Advances in Signal Processing*, 2003.

31. Q. Hong and L. Zhao, "Novel circuit designs of memristor synapse and neuron," *Neurocomputing*, 2019.

32. A. Sengupta and R. Wang, "Going deeper in spiking neural networks: Vgg and residual architectures," *Brain Sciences*, 2019.

33. C. Lee and K. Roy, "Enabling spike-based backpropagation in state-of-the-art deep neural network architectures," *arXiv preprint* arXiv:1903.06379, 2019.

34. E. Hunsberger and C. Eliasmith, "Training spiking deep networks for neuromorphic hardware," arXiv:1611.05141v1, 2016.

35. E. Hunsberger and C. Eliasmith, "Spiking deep convolutional neural networks for energy-efficient object recognition," *International Journal of Computer Vision*, 2016.

36. T. Wang, G. Yang, P. Chen, Z. Xu, M. Jiang, and Q. Ye, "A survey of applications of deep learning in radio signal modulation recognition," 12 2022.

37. F. Shi, Z. Hu, C. Yue, and Z. Shen, "Combining neural networks for modulation recognition," *Digital Signal Processing*, vol. 120, p. 103264, 1 2022.

38. U. Madhow, "Introduction to communication systems," 2014.

39. D. T. Kawamoto and R. W. Mcgwier, "Rigorous moment-based automatic modulation classification," 2016.

40. C.-F. Teng, C.-C. Liao, C.-H. Chen, and A.-Y. Wu, "Polar feature based deep architectures for automatic modulation classification considering channel fading," 2018.

41. Z. Wu, S. Zhou, Z. Yin, B. Ma, and Z. Yang, "Robust automatic modulation classification under varying noise conditions," *IEEE Access*, vol. 5, pp. 19733–19741, 8 2017.

42. F. Liu, Z. Zhang, and R. Zhou, "Automatic modulation recognition based on cnn and gru," 2022.

43. X. Hao, Z. Xia, M. Jiang, Q. Ye, and G. Yang, "Radio signal modulation recognition method based on deep learning model pruning," *Applied Sciences (Switzerland)*, vol. 12, 10 2022.

44. G. Wang, B. Li, T. Zhang, and S. Zhang, "A network combining a transformer and a convolutional neural network for remote sensing image change detection," *Remote Sensing*, vol. 14, 5 2022.

45. X. Xie, G. Yang, M. Jiang, Q. Ye, and C. F. Yang, "A kind of wireless modulation recognition method based on densenet and blstm," *IEEE Access*, vol. 9, pp. 125706–125713, 2021.

46. N. Wang, Y. Liu, L. Ma, Y. Yang, and H. Wang, "Multidimensional cnn-lstm network for automatic modulation classification," *Electronics (Switzerland)*, vol. 10, 7 2021.

47. W.-S. Hu, H.-C. Li, L. Pan, W. Li, R. Tao, and Q. Du, "Spatial-spectral feature extraction via deep convlstm neural networks for hyperspectral image classification," 5 2019.

48. W. Hu, H. Li, L. Pan, and Q. Du, "Spatial-spectral feature extraction via deep convlstm neural networks for hyperspectral image classification," *IEEE Trans. Geosci. Remote Sens*, vol. 58, pp. 4237–4250, 2020.

49. K. Jiang, J. Zhang, H. Wu, A. Wang, and Y. Iwahori, "A novel digital modulation recognition algorithm based on deep convolutional neural network," *Applied Sciences (Switzerland)*, vol. 10, 2 2020.

50. Y. Zheng, Y. Ma, and C. Tian, "Tmrn-glu: A transformer-based automatic classification recognition network improved by gate linear unit," *Electronics (Switzerland)*, vol. 11, 5 2022.

51. F. Tian, L. Wang, and M. Xia, "Signals recognition by cnn based on attention mechanism," *Electronics (Switzerland)*, vol. 11, 7 2022.

52. K. Liao, Y. Zhao, J. Gu, Y. Zhang, and Y. Zhong, "Sequential convolutional recurrent neural networks for fast automatic modulation classification," *IEEE Access*, vol. 9, pp. 27182–27188, 2021.

53. T. Huynh and T. Nguyen, "Automatic modulation classification: A deep architecture survey," *IEEE Access*, vol. 9, pp. 142950–142971, 2021.

54. O. S. Mossad, M. Elnainay, and M. Torki, "Deep convolutional neural network with multi-task learning scheme for modulations recognition," pp. 1644–1649, Institute of Electrical and Electronics Engineers Inc., 6 2019.

55. F. Zhang, C. Luo, J. Xu, Y. Luo, and F. Zheng, "Deep learning based automatic modulation recognition: Models, datasets, and challenges," 7 2022.

56. H. Han, Z. Ren, L. Li, and Z. Zhu, "Automatic modulation classification based on deep feature fusion for high noise level and large dynamic input," *Sensors*, vol. 21, pp. 1–13, 3 2021.

57. S. Ansari, K. A. Alnajjar, M. Saad, S. Abdallah, and A. A. El-Moursy, "Automatic digital modulation recognition based on genetic-algorithm-optimized machine learning models," *IEEE Access*, vol. 10, pp. 50265–50277, 2022.

58. A. Jagannath and J. Jagannath, "Multi-task learning approach for automatic modulation and wireless signal classification," 1 2021.

59. M. M. Elsagheer and S. M. Ramzy, "A hybrid model for automatic modulation classification based on residual neural networks and long short term memory," *Alexandria Engineering Journal*, vol. 67, pp. 117–128, 3 2023.

60. J. Ou, P. M. Ferreira, P. M. A. Ferreira, and M. Tunable, "A tunable morris-lecar spiking neuron in cmos," 2023.

61. E. J. Basham and D. W. Parent, "A neuromorphic quadratic, integrate, and fire silicon neuron with adaptive gain," pp. 1771–1776, IEEE, 7 2018.

62. L. Zhang, "Building logistic spiking neuron models using analytical approach," *IEEE Access*, vol. 7, pp. 80443–80452, 2019.

63. N. Brunel and P. E. Latham, "Firing rate of the noisy quadratic integrate-and-fire neuron," *Neural Compute*, 2003.

64. S. Song, B. Jeon, S. Hwang, M. H. Baek, J. H. Lee, and B. G. Park, "Integrate-and-fire neuron circuit with synaptic off-current blocking operation," *IEEE Access*, vol. 9, pp. 127841–127851, 2021.

65. G. Indiveri and T. K. Horiuchi, "Frontiers in neuromorphic engineering," *Frontiers in Neuroscience*, 2011.

66. C. D. Schuman, T. E. Potok, R. M. Patton, J. D. Birdwell, M. E. Dean, G. S. Rose, and J. S. Plank, "A survey of neuromorphic computing and neural networks in hardware," 5 2017.

67. G. Indiveri, F. Stefanini, and E. Chicca, "Spike-based learning with a generalized integrate and fire silicon neuron," pp. 1951–1954, 2010.

68. X. Wu, V. Saxena, K. Zhu, and S. Balagopal, "A cmos spiking neuron for brain-inspired neural networks with resistive synapses and in situ learning," *IEEE Transactions on Circuits and Systems II: Express Briefs*, vol. 62, pp. 1088–1092, 11 2015.

69. M. W. Kwon, K. Park, and B. G. Park, "Low-power adaptive integrate-and-fire neuron circuit using positive feedback fet co-integrated with cmos," *IEEE Access*, vol. 9, pp. 159925–159932, 2021.

70. A. K. Shah, E. S. Cho, J. Park, H. Shin, and S. Cho, "A compact integrate-and-fire neuron circuit embedding operational transconductance amplifier for fidelity enhancement," *IEEE Access*, vol. 11, pp. 53932–53938, 2023.

71. B. Liu, S. Konduri, R. Minnich, and J. Frenzel, "Implementation of pulsed neural networks in cmos vlsi technology," *International Conference on Signal Processing, Robotics and Automation*, 2005.

72. E. Chicca and F. Stefanini, "Neuromorphic electronic circuits for building autonomous cognitive systems," *Proceedings of the IEEE*, 2014.

73. C. Mead, "Neuromorphic electronic systems," *Proceedings of the IEEE*, 1990.

74. B. Han and A. Sengupta, "On the energy benefits of spiking deep neural networks: A case study," *International Joint Conference on Neural Networks*, 2016.

75. D. Liu and Y. Chai, "Low-power computing with neuromorphic engineering," *Advanced Intelligent Systems*, 2020.

76. R. A. M. Iannella and G. Indiveri, "Spike-based synaptic plasticity in silicon: Design, implementation, application, and challenges," *Proceedings of the IEEE*, 2014.

77. P. B. A. Tete and A. Keskar, "Design of dynamic synapse circuits with vlsi design approach," *International Conference on Emerging Trends in Engineering and Technology*, 2010.

78. K. Yamazaki and D. Bulsara, "Spiking neural networks and their applications: A review," *Brain Sciences*, 2022.

79. D. Christensen and R. Dittmann, "Roadmap on neuromorphic computing and engineering," *Neuromorphic Computing and Engineering*, 2022.

80. A. Aamir and P. Muller, "A mixed-signal structured adex neuron for accelerated neuromorphic cores," *IEEE Transactions on Biomedical Circuits and Systems*, 2018.

81. D. Pradhan and A. Sreedevi, "Analysis of the dynamic behaviour of a single hodgkin-huxley neuron model," *International Conference on Emerging Research in Electronics, Computer Science and Technology*, 2015.

82. L. Abbott, "Lapicque's introduction of the integrate-and-fire model neuron," *Brain Research Bulletin*, 1999.

83. S. K. Vohra, S. A. Thomas, M. Sakare, and D. M. Das, "Cmos circuit implementation of spiking neural network for pattern recognition using on-chip unsupervised stdp learning," 4 2022.

84. L. Z. Q. Hong and X. Wang, "Novel circuit designs of memristor synapse and neuron," *Neurocomputing*, 2018.

85. S. D. F. Zohora and A. Rashid, "Memristor-cmos hybrid implementation of leaky integrate and fire neuron model," *International Conference on Electrical, Computer and Communication Engineering*, 2019.

86. M. M. C. Yakopcic and D. Palmer, "Memristor-based neuron circuit and method for applying learning algorithm in spice," *Electronics Letters*, 2014.

87. E. C. G. Indiveri and R. Douglas, "A vlsi array of low-power spiking neurons and bistable synapses with spike-timing dependent plasticity," *IEEE Transactions on Neural Networks*, 2006.

88. G. Maranhão and J. G. Guimarães, "Low-power hybrid memristor-cmos spiking neuromorphic stdp learning system," *IET Circuits, Devices and Systems*, vol. 15, pp. 237–250, 5 2021.

89. E. M. Izhikevich, "Hybrid spiking models," 11 2010.

90. A. Basu, C. Frenkel, L. Deng, and X. Zhang, "Spiking neural network integrated circuits: A review of trends and future directions,"

91. B. Rajendran, A. Sebastian, M. Schmuker, N. Srinivasa, and E. Eleftheriou, "Low-power neuromorphic hardware for signal processing applications," 1 2019.

92. C. Mead, *Analog VLSI Implementation of Neural Systems*. Kluwer Academic Publishers, 1989.

93. N. Qiao, H. Mostafa, F. Corradi, M. Osswald, F. Stefanini, D. Sumislawska, and G. Indiveri, "A reconfigurable on-line learning spiking neuromorphic processor comprising 256 neurons and 128k synapses," *Frontiers in Neuroscience*, vol. 9, 2015.

94. J. Park and S. Cho, "A compact $integrate - and - fire$ neuron circuit embedding operational transconductance amplifier for fidelity enhancement," *IEEE Access*, 2023.

95. E. J. Basham and D. W. Parent, "An analog circuit implementation of a quadratic integrate and fire neuron," *Proceedings of the 31st Annual International Conference of the IEEE Engineering in Medicine and Biology Society: Engineering the Future of Biomedicine, EMBC 2009*, pp. 741–744, 2009.

96. A. Nowbahari, L. Marchetti, and M. Azadmehr, "Subthreshold modeling of a tunable cmos schmitt trigger," *IEEE Access*, vol. 11, pp. 10977–10984, 2023.

97. D. Hajtáš and D. Duračková, "Switched capacitor-based implementation of integrate-and-fire neural networks," *Journal of Electrical Engineering*, vol. 54, pp. 208–212, 2003.

98. A. V. Schaik, C. Jin, A. McEwan, and T. J. Hamilton, "A log-domain implementation of the izhikevich neuron model," *ISCAS 2010 - 2010 IEEE International Symposium on Circuits and Systems: Nano-Bio Circuit Fabrics and Systems*, pp. 4253–4256, 2010.

99. S. Millner, A. Grübl, G. Grübl, K. Meier, J. Schemmel, and M.-O. Schwartz, "A vlsi implementation of the adaptive exponential integrate-and-fire neuron model," *Kirchhoff-Institut fur Physik*, 2010.

100. M. Kimura, Y. Shibayama, and Y. Nakashima, "Neuromorphic chip integrated with a large-scale integration circuit and amorphous-metal-oxide semiconductor thin-film synapse devices," *Scientific Reports*, vol. 12, 12 2022.

101. S. R. Schultz and M. A. Jabri, "Analogue vlsi 'integrate-and-fire' neuron with frequency adaptation," *Electronics Letters*, vol. 31, pp. 1357–1358, 8 1995.

102. J. J. Lee, J. Park, M. W. Kwon, S. Hwang, H. Kim, and B. G. Park, "Integrated neuron circuit for implementing neuromorphic system with synaptic device," *Solid-State Electronics*, vol. 140, pp. 34–40, 2 2018.

103. J. G. T. Koickal and T. Pearce, "Analog vlsi circuit implementation of an adaptive neuromorphic olfocation chip," *IEEE Transactions on Circuits and Systems*, 2007.

104. K. Nalliboyina and S. Ramachandran, "An energy-efficient hybrid cmos spiking neuron circuit design with a memristive based novel t-type artificial synapse," *AEU - International Journal of Electronics and Communications*, vol. 173, 1 2024.

105. J. V. Arthur and K. A. Boahen, "Silicon-neuron design: A dynamical systems approach," *IEEE Transactions on Circuits and Systems I: Regular Papers*, vol. 58, pp. 1034–1043, 2011.

106. W. Gerstner, W. Kistler, R. Naud, and L. Paninski, *Neuronal Dynamics-From Single Neurons to Networks and Models of Cognition*. Cambridge University Press, 2014.

107. Y. Li, X. Cui, Y. Zhou, and Y. Li, "A comparative study on the performance and security evaluation of spiking neural networks," *IEEE Access*, vol. 10, pp. 117572–117581, 2022.

108. S. Yu, "Neuro-inspired computing with emerging nonvolatile memorys," *Proceedings of the IEEE*, vol. 106, pp. 260–285, 2 2018.

109. A. Hazan and E. E. Tsur, "Neuromorphic analog implementation of neural engineering framework-inspired spiking neuron for high-dimensional representation," *Frontiers in Neuroscience*, vol. 15, 2 2021.

110. F. Danneville, C. Loyez, K. Carpentier, I. Sourikopoulos, E. Mercier, and A. Cappy, "Solid-state electronics solid state electronics a sub-35 pw axon-hillock artificial neuron circuit," 2018.

111. J. H. B. Wijekoon and P. Dudek, "Spiking and bursting firing patterns of a compact vlsi cortical neuron circuit," pp. 1332–1337, IEEE, 8 2007.

112. I. Sourikopoulos, S. Hedayat, C. Loyez, F. Danneville, V. Hoel, E. Mercier, and A. Cappy, "A 4-fj/spike artificial neuron in 65 nm cmos technology," *Frontiers in Neuroscience*, vol. 11, 3 2017.

113. G. Indiveri and S.-C. Liu, "Memory and information processing in neuromorphic systems," 6 2015.

114. M. J. Rozenberg, O. Schneegans, and P. Stoliar, "An ultra-compact leaky-integrate-and-fire model for building spiking neural networks," *Scientific reports*, vol. 9, p. 11123, 7 2019.

115. V. Kornijcuk, H. Lim, J. Y. Seok, G. Kim, S. K. Kim, I. Kim, B. J. Choi, and D. S. Jeong, "Leaky integrate-and-fire neuron circuit based on floating-gate integrator," *Frontiers in Neuroscience*, vol. 10, 2016.

116. V. Bandeira, V. L. Costa, G. Bontorin, R. Reis, R. R. Low, V. V. Bandeira, and R. A. L. Reis, "Low latency fpga implementation of izhikevich-neuron model," pp. 210–217, 2015.

117. E. J. Basham and D. W. Parent, "Compact digital implementation of a quadratic integrate-and-fire neuron," pp. 3543–3548, 2012.

118. W. C. Wu, C. F. Yeh, A. J. White, C. T. Wang, Z. W. Yeh, C. C. Hsieh, R. S. Liu, K. T. Tang, and C. C. Lo, "Integer quadratic integrate-and-fire (iqif): A neuron model for digital neuromorphic systems," Institute of Electrical and Electronics Engineers Inc., 6 2021.

119. R. Hermida, M. Patrone, M. Pijuan, P. Monzón, and J. Oreggioni, "An analog circuit implementation of a huber-braun cold receptor neuron model," pp. 3376–3379, 2012.

120. J. Quan, Z. Liu, B. Li, and J. Luo, "Ultra-low-power compact neuron circuit with tunable spiking frequency and high robustness in 22 nm fdsoi," *Electronics (Switzerland)*, vol. 12, 6 2023.

121. T. Hamilton, "Compact and energy efficient neuron with tunable spiking neuron in 22-nm fdsoi," *TechRxiv*, 2021.
122. A. S. Alkabaa, O. Taylan, M. T. Yilmaz, E. Nazemi, and E. M. Kalmoun, "An investigation on spiking neural networks based on the izhikevich neuronal model: Spiking processing and hardware approach," *Mathematics*, vol. 10, 2 2022.
123. N. G. S, D. P. K. M. B, D. Mahesh, and M. Subramanyam, "Comparative research of neuron circuits," *International Journal for Research in Applied Science and Engineering Technology*, vol. 10, pp. 4121–4126, 7 2022.
124. L. Garaffa, A. Aljuffri, C. Reinbrecht, and J. Sepúlveda, "Revealing the secrets of spiking neural networks: The case of izhikevich neuron," pp. 514–518, Institute of Electrical and Electronics Engineers Inc., 2021.
125. V. Rangan, A. Ghosh, V. Aparin, and G. Cauwenberghs, "A subthreshold a vlsi implementation of the izhikevich simple neuron model," pp. 4164–4167, 2010.
126. E. M. Izhikevich, "Simple model of spiking neurons," 11 2003.
127. A. K. Arifuzzman, M. S. Islam, and M. R. Haider, "A neuron model based ultralow current sensor system for bioapplications," *Journal of Sensors*, vol. 2016, 2016.
128. T. Asai, Y. Kanazawa, and Y. Amemiya, "A subthreshold mos neuron circuit based on the volterra system," *IEEE Transactions on Neural Networks*, vol. 14, pp. 1308–1312, 9 2003.
129. V. Varshavsky, V. Marakhovsky, and H. Saito, "Cmos implementation of an artificial neuron training on logical threshold functions," 2009.
130. Q. Ma and M. R. Haider, "A silicon neuron based biopotential amplifier for biomedical applications," IEEE Computer Society, 2014.
131. K. Nakada and H. Hayashi, "Analog vlsi implementation of resonate-and-fire neuron," 2006.
132. K. Srinivasan and G. Cowan, "Subthreshold CMOS Implementation of the Izhikevich Neuron Model," International Symposium on Circuits and Systems, 2022.